John T King

Guide to Baltimore and Ohio Railroad

John T King

Guide to Baltimore and Ohio Railroad

ISBN/EAN: 9783337419578

Printed in Europe, USA, Canada, Australia, Japan

Cover: Foto ©Lupo / pixelio.de

More available books at **www.hansebooks.com**

THE
Baltimore & Ohio R
The Great National

CONNECTING ALL PARTS O

South and

WITH

ALL PARTS OF TH

GUID

To

BALTIMORE & OHIO

JOHN T. KING

1,000 Copies per Month of this Guide
the Hotels, News Depots, Railw
Passenger Trains of Baltim

Procure your Tick
Baltimore Street and
Baltimore, and at 485 P
enue, and Depot Corner
enue and C Sts., Washin

THOS
Passenger A

70 Miles S

VIA BALTIM

TO ALL POINTS

West and Sou

Than the Pennsylvan

INDEX.

	PAGE.
Time Tables Baltimore & Ohio Railroad	1
Connections of Baltimore & Ohio Railroad	7
Advertisements of Baltimore & Ohio Railroad	9
Bremen Steamship Line	16
Liverpool Steamship Line	16
Hotel Advertisements	17
General Advertisements	20
History of Baltimore & Ohio Railroad	24
Valley Railroad	31
Maryland	34
Historic Homes of the Valley of the Shenandoah	59
Advertising Rates	3d page of cover.

MAIN STEM BALTIMORE & OHIO RAILROAD.

J. W. GARRETT, President. JNO. KING, JR., President *pro tem.*
THO'S R. SHARP, Master Transp'tion. L. M. COLE, General Ticket Agent.
SIDNEY B. JONES, General Passenger Agent, Cincinnati, Ohio.

Acc.	Exp.	Exp.	Mail	Acc.	STATIONS.	Exp.	Mail	Exp.	Acc.	Acc.
P. M.	P. M.	A. M.	A. M.	P. M.	*leave]* [*arrive*	M.	P. M.	P. M.	A. M.	A. M.
4.10	6*00	6*45	8.00	5.00	**Baltimore.**	12.00	5.10	10.40	10.40	8.20
4.30	6.13	7.01	8.20	5.20	...Washington Junction...	11.41	4.50	10.22	10.20	7.50
......	7.20	7.50			......**Washington.**	10.40		9.30	......	......
4.48			8.37	5.38	........Ellicott's Mills........		4.33		10.03	7.33
5.03			8.51	5.53	...........Elysville...........		4.19		9.49	7.19
5.24			9.11	6.14	..........Marriottsville..........		3.58		9.29	7.00
5.36			9.23	6.27	...........Sykesville...........		3 45		9.16	6.47
6.12			9.58	7.00	...........Mount Airy...........		3.12		8.44	6.15
6.35			10.19	7.21	...........Monrovia...........		2.49		8.22	5.54
7.03			10.46	7.43	.Frederick Junction.		2.25		8.00	5.32
7.34	9.30	9.55	11.19		.......Point of Rocks.......	8.35	1.47	7.35	7.22	
8.02	9.56	10.18	11.49		HagerstownJunct'n	8.11	1.18	7.11	6.52	
8.05			11.57		......Sandy Hook......		1.14		6.48	
8.08	10.04	10.26	12.06		....Harper's Ferry.....	8.04	1.11	7.05	6.45	
......			12.24		..........Duffield's..........		12.49		A. M.	
			12.39		..........Kerneysville..........		12.35			
	10.55	11.15	1.20		..........Martinsburg..........	7.20	12.15	6.20	Winc. Acc.	
			1.41		.........North Mountain.........		11.37			
			2.11		.........Sleepy Creek.........		11.09			
			2.26		...........Hancock...........		10 55			
Cumb Acc. 12.00	12.13	2.44			.........Sir John's Run.........	5.57	10.38	5.17	Cumb Acc.	
			4.11		.........Little Cacapon.........		9.20			
			4.28		......Green Spring Run......		9.03			
A. M.			4.47		......Patterson's Creek......		8.45		P. M.	
7.30	2.05	2.17	5.15		..........**Cumberland.**..........	4.15	8.23	3.38	8.20	.....
7.47			5.33		..........Brady's Mill..........		7.55		7.59	
8.21			6.17		.........New Creek.........		7.12		7.10	
8.35	3.05	3.20	6.30		...........Piedmont...........	3.10	7.00	2 20	6.55	
9.02					...........Frankville...........				6.25	
9.21					...........Swanton...........				6.06	
9.31					...........Altamont†...........				5.57	
9.41	4.14	4.23			.......Deer Park...........	2 06		1.17	5.48	
9.58	4.28	4.37			..........**Oakland**..........	1.52		1.03	5.30	
10.26					.....Cranberry Summit....				5.00	
11.03	5.23	5.32			........Rowlesburg........	12.53		12.04	4.12	
11.41					..........Tunnelton..........				3.49	
12.23					..........Thornton..........				3.06	
12.58	7.10	7.05			..........**Grafton**..........	11.40		10.35	2.50	
1.03					..........Fetterman..........					
1.53					..........Benton's Ferry..........					
2.07	8.00	7.55			..........**Fairmont**..........	10.20		9.36	1.42	
2.37					..........Farmington					
2.57	8.47	8.33			..........Mannington..........	9.34		8.47	1.04	
3.29					..........Burton..........					
3.49					..........Littleton..........					
4.29	10.10	10.01			..........Cameron..........	8.10		7.22	11.47	
5.02					.Roseby's Rock.......					
5.29	10.49	10.42			.... Moundsville	7.31		6.37	11.10	
5.50	11.15	11.00			..........Benwood..........	7.15		6.20	10.40	
6.15	11.40	11.30			..........**Wheeling**..........	6*40		5*45	10.15	
P. M.	A. M.	P. M.			*arrive]* [*leave*	P. M.		A. M.	A. M.	

* Run daily, including Sundays. † 2,720 feet above tide-water.
Sundays, from Wheeling at 5.45 A. M., 6.40 P. M.—from Balt. 6.45 A. M., 6.00 P. M.

EAST & WEST VIA METROPOLITAN BRANCH.

P. M.	P. M.	A. M.	leave] [arrive	A. M.	M.	P. M.
* 6.00	....	* 6.45	...............Baltimore.........	9.10	12.00	10.40
7.30	4.00	8.00	...............Washington.......	8.30	10.40	9.25
....	4.05	....	...... Metropolitan Junction...........	8.24	....	....
....	4.14	....	........Terra Cotta............	8.14	....	....
....	4.23	....	...........Silver Spring.............	7.59	....	....
....	4.35	....	.... Knowles......	7.47	....	....
8.17	4.51	8.46	...............Rockville.................	7.31	9.56	8.43
....	5.07	....	...... Gaithersburg................	7.16	9.43	....
....	5.22	....	...... Germantown	7.01	....	....
....	5.32	....	...............Boyd's	6.50	....	....
....	5.44	....	..........Barnesville....	6.38	9.03	....
....	5.52	....	... Dickerson's.............	6.30	....	....
....	6.02	....	............Tuscarora............	6.20	....	....
9.30	6.15	9.55	... Point of Rocks..............	6.10	8.35	7.35
....	....	10.18	............Hagerstown Junction.............	....	8.11	7.11
10.04	....	10.26	..........Harper's Ferry...........:	....	8.04	7.05
10.55	....	11.15	............Martinsburg	....	7.20	6.20
12.00	....	12.13	............Sir John's Run..............	....	5.57	5.17
2.05	....	2.17	...Cumberland.............	....	4.15	3.38
3.10	....	3.20	............Piedmont...............	....	3.10	2.20 ·
4.14	....	4.23	....Deer Park.............	....	2.06	1.17
4.28	....	4.37	............Oakland.............	....	1.52	1.03
5.23	P. M.	5.32	.... Rowlesburg...............	P. M.	....	12.04
7.00	2.50	7.05	.............Grafton..............	12.05	11.40	10.50
7.45	3.47	7.51	..............Clarksburg	11.05	10.41	9.54
....	5.01	....	..............West Union.......	9.49	....	....
8.43	5.10	8.53	.... Central..............	9.40	9.36	8.48
....	5.54	....	.... Ellenboro............	8.49	....	....
9.57	6.42	10.00	....L. F. Junction.................	8.00	8.25	7.43
10.45	7.35	10.45	..................Parkersburg..................	7.06	7.45	7.00
A. M.	P. M.	P. M.		A. M.	P. M.	A. M.
7.10	....	7.05	...............Grafton.............	....	11.20	10.35
10.10	....	10.01	..............Cameron.......	....	8.10	7.22
10.49	....	10.42	............Moundsville..............	....	7.31	6.37
11.15	....	11.00	......:........ Benwood...............	....	7.15	6.20
11.40	....	11.30	..........:.......Wheeling..............	....	* 6.40	* 5.45

PITTSBURG, WASHINGTON AND BALTIMORE R. R.

CONNELLSVILLE ROUTE.

Leave Baltimore.....5 45 A. M. 6.00 P. M	Leave Pittsburg... 8.00 A. M. 8.30 P. M.
" Washington..8.00 A. M. 7.45 P. M.	" Cumberl'd.. 3.18 P. M. 4.07 A. M.
" Cumberland..1.57 P. M. 2.00 A. M.	" Washingt'n 9.25 P. M. 10.40 A. M.
Arrive Pittsburg.....9.40 P. M. 9.00 A. M.	Arrive Baltimore..10.40 P. M. 12.00 M.

WASHINGTON COUNTY DIVISION B. & O. R. R.

P. M.	A. M.	A. M.	A. M.	Leave.	Arrive.	A. M.	A. M.	P. M.	P. M.
5.50	10.25	9.00	6.50	.....Hagerstown.....		9.35	11.40	2.35	9.20
6.07	10.42	9.18	7.07	Breathed's Station		9.18	11.23	2.18	9.03
6.19	10.53	9.30	7.19	Keedysville		9.01	11.11	2.06	8.51
6.25	11.05	9.36	7.25	Eakle's Mill		8.55	11.05	2.00	8.45
6.33	11.12	9.44	7.33	Rhororsville		8.47	10.57	1.52	8.37
6 37	11.15	9.48	7.37	Beeler's Summit		8.43	10.53	1.48	8.33
6.48	11.25	9.59	7.48	Bartholow's		8.32	10.42	1.37	8.22
7.00	11.35	10.10	8.00	Hagerstown Junction		8.20	10.30	1.25	8.10
P. M.	A. M.	A. M.	A. M.	Arrive.	Leave.	A. M.	A. M.	P. M.	P. M.

FREDERICK AND ELLICOTT CITY ACCOM.

A. M.	P. M.	P. M.	P. M.	Leave, Sundays Excepted,	A. M.	A. M.	P. M.	P. M.
8.00	2.30	5.00	1.20	Baltimore	7.10	8.20	3.20	5.10
8.20	2.50	5.20	1.50	Washington Junction	6.50	7.50	2.50	4.50
8.37	3.10	5.38	2.10	Ellicott City	6.30	7.33	2.30	4.33
8.51		5.53		Elysville		7.19		4.19
9.11		6.14		Marriotsville		7.00		3.58
9.23		6.27		Sykesville		6.47		3.45
9.58		7.00		Mount Airy		6.15		3.12
10.19		7.21		Monrovia		5.54		2.49
10.45		7.43		Frederick Junction		5.32		2.25
11.00		7.55		Frederick		5.20		

Winchester, Potomac and Harrisonburg Division.

P. M.	A. M.	Leave.	Arrive.	P. M.	A. M.
8.23	10.40	Harper's Ferry	6.35		6.22
8.27	10.44	Shenandoah	6.31		6.18
8.39	10.56	Halltown	6.18		6.06
8.49	11.07	Charlestown	6.08		5.55
8.59	11.18	Cameron	5.55		5.44
9.09	11.29	Summit Point	5.44		5.33
9.22	11.42	Wadesville	5.30		5.21
9.35	11.56	Stephenson's	5.16		5.09
9.45	P. M.	} Winchester {	5.04		5.00
	12.13				A. M.
	12.27	Kernstown	4.45		
	12.32	Bartonville	4.40		
	12.37	Newtown	4.35		
	12.43	Vaucluse	4.29		
	12.49	Middletown	4.23		
	12.55	Cedar Creek	4.17		
	1.03	Capon Road	4.00		
	1.06	Strasburg Junction	4 06		
		Strasburg			
	3.30	Harrisonburg	1.40		

STRAITSVILLE DIVISION.

P. M.	A. M.		P. M.	P. M.
4.15	7.00	Leave... Newark ....Arrive.	12.55	9.20
4.35	7.20	National Road	12.28	8.55
4.42	7.27	Avondale	12.15	8.48
4.48	7.33	Thornport	12.03	8.42
5.04	7.50	Glenford	11.40	8.26
5.24	8.09	Somerset	11.13	8.02
5.34	8.19	Wellan's	11.00	7.50
5.50	8.35	Junction City	10.37	7.33
6.07	8.50	Bristol	10.20	7.19
6.21	9.02	McCuneville	10.07	7.07
6.30	9.10	Arrive..Shawnee..............Leave.	10.00	7.00

WASHINGTON BRANCH
AND
NEW YORK AND WASHINGTON AIR LINE.

Going North.

	A.M.	A.M.	A.M.	A.M.	P.M.	P.M.	P.M.	P.M.	P.M.	P.M.	P.M.
WASHINGTON	5.00	6.45	8*00	10*50	12.00	2*00	3.30	4*45	5.45	6.45	9*3
Metropolitan Junc.	5.04	6.49	8.45		12.04	2.04	3.45	4*49			7.49	9*40
Bladensburg	5.17	7.03	8.49		12.18	2.18	3.49	5.03			8.01	9.56
Beltsville	5.33	7.30	9.03		12.35	2.35	4.03	5.20			8.14	10.07
Contee's	5.41	7.29	9.20		12.44	2.44	4.20	5.29			8.21	
Laurel	5.46	7.34	9.29		12.49	2.49	4.29	5.34	3.55	7.29	8.26	
Savage	5.51	7.39	9.31		12.54	2.54	4.34	5.39		7.34	8.31	
ANNAPOLIS JUNC'N	†7.43	9.30		12.58	2.58	4.39	‡5.43		7.43	8.34	10.23	
Jessup's Cut	5.59	7.48	9.42		1.03	3.03	4.43	5.48		7.48	8.39	
Dorsey's Cut	6.03	7.53	9.48		1.08	3.08	4.53	5.53		7.53	8.44	
Hanover	6.08	7.58	9.58		1.13	3.13	4.58	5.58		7.58	8.49	
Elk Ridge Landing	6.13	8.03	10.03		1.18	3.18	5.03	6.03		8.03	8.51	
Relay	6.16	8.06	10.06		1.21	3.21	5.06	6.06		8.06	8.57	10.22
BALTIMORE	6.35	8.26	9.10	10.25	1.40	3.40	5.25	6.25		8.25	9.15	10.42
PHILADELPHIA	11.45		1.30			6.55				2.50		10.42
NEW YORK		4.30		10.50		10.08				11.00		6.35

Going South.

	P.M.	A.M.	A.M.	A.M.	A.M.	P.M.	P.M.	P.M.	P.M.	P.M.	P.M.
NEW YORK	9.00							9.00		3.15	6.00
PHILADELPHIA	12.40					9.00			4.00	6.00	
BALTIMORE	4*20	5*20	8.45	10.00	11.00	12.15	3.30	4.00	4*45	5.45	6.45	9*3	9*40
Relay House	5*37	7.40		11.00		3.45	4.17	4.47	6.00	6.18		
Elk Ridge Landing	4.38	5.37	7.01	7.56	10.20		3.48	4.17	4.47	6.00	6.18	10.30
Hanover Cut			7.08		11.18		3.51	4.30		5.06	6.08	
Dorsey's Cut			7.11	8.03	11.21	1.21	3.57	4.32		5.13	6.13	
Jessup's Cut			7.23	9.11	11.27	1.27	4.03	4.36		5.28	6.36	
ANNAPOLIS JUNC'N	4.36		†7.33	9.16	11.33	1.33	4.08	†5.06	5.37	6.40	
Savage			7.37	9.21	11.38	1.38	4.13	4.17	5.47		
Laurel			7.42	9.29	11.43	1.43	4.17	4.22	5.50	7.14	9.32
Contee's			7.47	9.34	11.47	1.47	4.22	4.48	5.30	7.24	9.24
Beltsville			7.56	9.39	11.52	1.52	4.27		6.00	7.31	9.32
Bladensburg			8.12	9.46	12.06	2.06	4.36	5.28	6.08	7.44	9.47
Metropolitan Junc.			8.26	9.59	12.22	2.22	4.52		6.13	7.56	9.56
WASHINGTON	5.40	6.30	8.30	10.11	11.10	2.36	5.06	5.15	6.36	8.00	9.56
		7.50	8.40	10.15	11.10	2.40	5.10	5.15	6.40	7.20	8.00	10.00	11.40

† Connect with trains for Annapolis. * Run Sundays. ‡ Leaves at 1.30 on Sundays.

PARKERSBURG DIVISION B. & O. R. R.

A.M.	P.M.	P.M.	LEAVE.	ARRIVE.	A.M.	P.M.	P.M.
7.00	2.50	7.05	.................GRAFTON..................		10.42	12.05	11.30
....	3.01	....	.... Webster			11.54	....
....	3.08	....	.... Simpson's			11.45	....
....	8.14	....	.... Flemington			11.38	....
....	8.33	....	....................Bridgeport...................			11.19	....
7.45	3.47	7.51	...................Clarksburg...................		9.54	11.05	10.41
....	3.59	....	...................Wilsonburg...................			10.52	....
....	4.10	....	..................Wolf's Summit..................			10.43	....
....	4.16	....	.... Brandy Gap			10.37	....
....	4.21	....	..................Cherry Gap..................			10.31	....
....	4.27	....	.... Salem			10.25	....
....	4.41	....	.... Long Run			10.11	....
....	4.55	....	.... Smithton			9.56	....
....	5.01	....	.... West Union....			9.49	....
8.48	5.10	8.53	....................Central....................		8.48	9.40	9.36
....	5.28	....	.... Toll Gate			9.23	....
....	5.40	....	.... Pennsborough			9.11	....
....	5.54	....	.... Ellenborough			8.49	....
....	6.08	....	.... Cornwallis..................			8.34	....
....	6.16	....	.... Cairo			8.26	....
....	6.25	....	...................Silver Run...................				....
....	6.36	....	.... Petroleum...................			8.06	....
9.57	6.42	10.00	.................L. F. Junction.................		7.43	8.00	8.25
....	6.45	....	.... Eaton's			7.57	....
....	6.56	....	.... Walker's			7.46	....
....	7.09	....	.... Kanawha			7.33	....
....	7.17	....	.... Claysville			7.24	....
....	7.25	....	.... Jackson's			7.15	....
10.45	7.35	10.45	..................PARKERSBURG..................		7.00	7.06	7.45
A.M.	.M.	P.M.	ARRIVE.	LEAVE.	A.M.	A.M.	P.M.

LAKE ERIE DIVISION B. & O. R. R.

A.M.	A.M.	P.M.	LEAVE.	ARRIVE.	P.M.	P.M.	P M.
11.45	7.35	5.30	.... SANDUSKY		6.25	10.25	4.20
12.10	7.57	6.25	.... Prout's		6.00	10.00	3.43
12.40	8.22	7.30	.... Monroeville		5.35	9.35	2.50
12.49	8.32	7.55	......................Pontiac......................		5.26	9.15	2.30
12 57	8.42	8.20	......................Havana......................		5.17	9.05	2.16
1.06	8.52	8.54	.... Centreton		5.06	8.54	2.00
1.19	9.09	9.20	.... New Haven		4.50	8.37	1.34
1.24	9.15	9.32	.... Plymouth		4.42	8.30	1.24
1.40	9.35	10.35	......................Shelby Junction......................		4.23	8.10	12.30
1.55	9.55	11.05	.... Spring Mill		4.07	7.32	11.15
2.08	10.10	11.30	.... MANSFIELD		3.52	7.36	10.48
2.31	10.38	12.25	.... Lexington		3.28	7.10	9.55
2.45	11.53	1.00	......................Belleville......................		3.13	6.54	9.26
2.57	11.07	1.35	.... Independence		2.57	6.39	8.55
3.10	11.24	2.10	.... Ankenytown		2.42	6.24	8.25
3.20	11.35	2.40	.... Frederick		2.33	6.13	8.02
3.26	11.53	3.10	.... Mount Vernon		2.17	5.55	7.26
3.48	12.06	3.38	.... Hunt's		2.03	5.41	6.59
4.01	12.21	4.05	.... Utica		1.48	5.26	6.30
4.11	12.32	4.25	.... Louisville		1.38	5.15	6.08
4.19	12.40	4.40	.... Vannatta's		1.30	5.06	5.53
4.35	12.55	5.10	.... NEWARK		1.15	4.50	5.20
P.M.	P.M.	A.M.	ARRIVE.	LEAVE.	P.M.	P.M.	A.M.

CENTRAL OHIO DIVISION B. & O. R. R.

A.M.	A.M.	P.M.	A.M.	Leave. Arrive.		A.M.	P.M.	P.M.
3.10	3.20	7.00	11*50	... Columbus ...	....	10.00	2.40	*5.50
3.20	3.31	7.12	....	..Alum Creek..	....	9.48	2.28	....
3.28	3.41	7.22	....	...Big Walnut...	....	9.38	2.18	....
3.30	3.43	7.25	....	... Taylor's....	....	9.36	2.15	....
3.34	3.48	7.30	....	..Black Lick...	....	9.31	2.09	....
3.42	3.57	7.40	....	....Summit....	....	9.22	2.00	....
3.47	4.02	7.47	....	...Columbia....	....	9.16	1.53	....
3.49	4.05	7.50	....	.. Pataskala....	....	9.13	1.50	5.13
3.58	4.15	8.00	....	.. Kirkersville..	....	9.02	1.40	....
4.08	4.27	8.12	....	.....Union.....	....	8.48	1.27	4.54
4.14	4.33	8.19	....	.... Granville...	....	8.42	1.21	....
4.23	4.50	8.30	1 10	.... Newark...	....	8.30	1.10	4.40
A.M.	5.06	8.47	....	...Clay Lick...	....	8.08	12.39	....
....	5.17	8.59	....	..Black Hand..	....	7.56	12.28	3.59
....	5.27	9.10	....	... Claypool's ...	....	7.45	12.18	....
....	5.36	9.20	....	.Pleasant Valley.	....	7.36	12.09	....
....	5.49	9.33	....	...Dillon's Falls.	P. M.	7.22	11.56	....
....	6.15	10.00	2.35	.. ZANESVILLE..	10.45	7.10	11 45	3.15
....	6 25	10.12	....	...Coal Dale....	10.32	A. M.	11.14	....
....	6.34	10.22	....	....Sonora....	10.22	....	11.05	....
....	6.51	10.42	3.12	....Norwich	10.04	....	10.46	2.17
....	7.01	10.52	3.21	....Concord	9.54	....	10.37	2.08
....	7.14	11.06	....	... Cassel's	9.41	....	10.24	....
....	7.25	11.18	3.45	.. CAMBRIDGE..	9.30	....	10.12	1.43
....	7.45	11.41	....	...Campbell's...	9.05	...	9.52	...
....	7.52	11.50	....	...Gibson's....	8.57	....	9.45	....
....	8.02	12 02	....	...Salesville ...	8.45	...	9.35	....
...	8.08	12.09	....	..Quaker City..	8.37	...	9.30	....
....	8.13	12.15	....	... Spencer's ...	8.30	....	9.25	12.57
....	8.30	12.35	4.55	...Barnesville...	8.10	....	9.08	12.40
....	8.42	12.48	....	.....Burton's...	7.58	....	8.57	....
....	8.50	12.56	....	..Burr's Mill...	7.51	....	8.50	....
....	8.55	1.02	5.20	....Belmont....	7.45	...	8.43	12.15
....	9.04	1.12	....	...Lewis' Mill...	7.35	....	8.33	...\
...	9.14	1.23	5.39	....Warnock ...	7.23	..	8.22	....
....	9.23	1.33	....	.... Glencoe	7.12	...	8.12	....
....	9.36	1.45	....	..Neff's Siding..	7.00	....	8.00	....
....	9.50	2.00	6.15	.... Bellaire ...	6.45	...	7.45	11 20
A.M.	A.M.	P.M.	Arrive. Leave.	P.M.		A.M.	A.M.

* Run Sundays.

BALTIMORE AND OHIO RAILROAD.

Main Line and Parkersburg Division.

JOHN W. GARRETT, President, Baltimore, Md.
JOHN KING, First Vice-President, Baltimore, Md.
WM. KEYSER, Second Vice-President, Baltimore, Md.
W. H. IJAMS, Secretary and Treasurer, Baltimore, Md.
WM. T. THELIN, Auditor, Baltimore, Md.
A. D. SMITH, Assistant Auditor, Columbus, Ohio.
THOMAS R. SHARP, Master of Transportation, Baltimore, Md.
JOHN L. WILSON, Master of Road, Baltimore, Md.
W. C. QUINCY, Gen. Superintendent Ohio Division, Columbus, Ohio.
L. M. COLE, General Ticket Agent, Baltimore, Md.
A. A. JOHNSON, Assistant General Ticket Agent, Baltimore, Md.
SIDNEY B. JONES, General Passenger Agent, Cincinnati, Ohio.
N. GUILFORD, General Freight Agent, Baltimore, Md.
G. B. SPRIGGS, General Agent, Columbus, Ohio.
C. A. CHIPLEY, Agent, 87 Washington street, Boston, Mass.
BENJ. WILLIAMS, Purchasing Agent, Baltimore, Md.
JOHN C. DAVIS, Master Mechanic, Baltimore, Md.
H. SIMDORN, Master Car Builder, Baltimore, Md.

CONNECTIONS.

(1) With Philadelphia, Wilmington & Baltimore R. R.; also with Line of Steamers to Europe, and with steamers from Baltimore to Norfolk and Portsmouth.
(2) With Washington Branch. (3) With Washington County R. R. Division. (4) With Winchester, Potomac and Strasburg Division. (5) With Stages for Berkeley and Bedford Springs. (6) With Connellsville Route. (6½) With Cumberland & Pennsylvania R. R. (7) With Parkersburg Division, over which, in connection with Marietta & Cincinnati R. R., through cars are run from Baltimore to Cincinnati, without change. (8) With Laurel Fork & Sand Hill R. R. (9) With Marietta & Cincinnati R. R. (10) With Ohio & Mississippi R. W.; also with Indianapolis, Cincinnati & Lafayette R. R.; with United States Mail Line of Steamers for Louisville and points on the Ohio River; also with the Louisville, Cincinnati & Lexington Short Line Route to Louisville, &c.

---

(1) With Philadelphia, Wilmington & Baltimore Railroad, for New York, Philadelphia, and all Eastern cities; with Steamers for Europe, and with Steamship Line for Norfolk and Portsmouth.
(2) Parkersburg Division diverges from Main Line.
(3) With Central and Ohio Division.
(4) With Cleveland & Pittsburg Railroad, and with Steamers from Wheeling to river ports.
(5) With Cincinnati & Muskingum Valley Railroad.
(6) With Pittsburg, Cincinnati & St. Louis Railway; also with Lake Erie Division of Baltimore & Ohio Railroad.
(7) With Little Miami, and Indianapolis and Chicago Division of Pittsburg, Cincinnati & St. Louis Railway, for Cincinnati, Indianapolis, Chicago, &c.; also with Main Line of Cleveland, Columbus, Cincinnati & Indianapolis Railway, and branch from Columbus to Springfield.
(8) With Cleveland, Mount Vernon & Delaware Railroad.
(9) With Pittsburg, Fort Wayne & Chicago Railway.
(10) With Cleveland, Columbus, Cincinnati & Indianapolis Railway.

NOTES ON RUNNING OF TRAINS.

WESTWARD.

Cincinnati Express—Leaving Baltimore 6.45 A. M. daily, connects with 9 P. M. train from New York, and runs through to Cincinnati without change.

St. Louis Express—Leaves Baltimore daily, 6.00 P. M. No change of cars from Baltimore to Cincinnati or Pittsburg.

☞ All other Westward-bound trains daily, except Sunday.

EASTWARD.

Cincinnati Express—Leaving Cincinnati 7.35 A. M., runs daily, except Sunday, arriving at Baltimore 8.50 A. M., and at New York 4.25 P. M.

St. Louis Express—Leaving Cincinnati 10.30 P. M., Parkersburg 7 A. M., runs daily, arriving at Baltimore 10.20 P. M., and makes close connection with express train, arriving at New York 6.30 A. M.

☞ All other Eastward-bound trains daily, except Sunday.

Ellicott's Mills Accommodation Trains leave Camden Station 1.20 P. M., arriving at Ellicott's Mills 2.10 P. M. Returning, leave Ellicott's Mills 2.30 P. M., arriving at Camden Station 3.20 P. M.

Winchester Accommodation Train leaves Baltimore 4.10 P. M., arriving at Winchester 5 A. M. Returning, leaves Winchester 9.40 P. M., arriving at Baltimore 10.40 A. M.

Comparative Distances to Baltimore and New York.

From Chicago, Ill.—To Baltimore, via Baltimore and Ohio railroad, 795 miles. To New York—Via New York Central railroad, 980 miles; via Erie railroad, 961 miles; via Pennsylvania railroad, 899 miles. Less to Baltimore than the average distance to New York, 152 miles.

From St. Louis, Mo.—To Baltimore, via Baltimore and Ohio railroad, 929 miles. To New York—Via New York Central railroad, 1,167 miles; via Erie railroad, 1,201 miles; via Pennsylvania railroad, 1,050 miles. Less to Baltimore than the average distance to New York, 210 miles.

From Louisville, Ky.—To Baltimore, via Baltimore and Ohio railroad, 696 miles. To New York—Via New York Central railroad, 989 miles; via Erie railroad, 987 miles; via Pennsylvania railroad, 851 miles. Less to Baltimore than the average distance to New York, 246 miles.

From Cincinnati, Ohio.—To Baltimore, via Baltimore and Ohio railroad, 589 miles. To New York—Via New York Central railroad, 882 miles; via Erie railroad, 861 miles; via Pennsylvania railroad, 744 miles. Less to Baltimore than the average distance to New York, 240 miles.

From Pittsburg, Pa.—To Baltimore by the Baltimore and Ohio railroad, 327 miles; to New York by the Pennsylvania railroad, 431 miles. Difference in favor of Baltimore, 104 miles, and from all points south of Baltimore 200 miles.

BALTIMORE & OHIO
RAILROAD.

The Quickest Time Ever Made
WEST!

Pullman's Palace Sleeping Cars on Night Trains.

☞ 15 Trains Daily
EACH WAY,

Between Baltimore and Washington!

Washington Passengers, Purchasing their Tickets at the New Office, No. 485 Pennsylvania Avenue;

or Baltimore Passengers, Purchasing their Tickets at the New Office, Corner Baltimore and Calvert Sts.

Can have their Baggage called for and Checked Through to destination, at their Residences or Hotels, before leaving for the Depot.

	From Baltimore.	From Washington.
TIME TO CINCINNATI,	23 hours 35 min.	22 hours 20 mins.
TIME TO LOUISVILLE,	28 " 55 "	27 " 45 "
TIME TO ST. LOUIS,	37 " 00 "	35 " 45 "

This Time is One Train in Advance of all Rival Lines !

THOS. R. SHARP, *Master of Transportation.*

L. M. COLE, *General Ticket Agent, Baltimore.*

SIDNEY B. JONES, *Gen'l Pass. Ag't, Cincinnati.*

1*

BALTIMORE AND OHIO RAILROAD.

CHANGE OF TIME.

On and after Sunday, November 2, 1873, trains will be run as follows:

The Cincinnati Express, via Metropolitan Branch Road, for Parkersburg, Cincinnati, Chicago and St. Louis, will leave daily at 6.45 A. M.

The St. Louis Express, with Pullman cars, for Pittsburg, Chicago and St. Louis, via Washington and Metropolitan Branch Road, will leave daily at 6.00 P. M.

The Mail Train, via Main Stem, for Piedmont, and all Local Stations east of that point will leave at 8.00 A. M. daily (except Sunday.)

The Winchester Accommodation, via Main Stem, will leave daily (except Sunday) at 4.10 P. M., stopping at all Stations.

The Frederick Accommodation, via Main Stem, will leave at 5.00 P. M. daily.

Ellicott's Mills Accommodation will leave daily (except Sunday) at 1.20 P. M.

FOR HAGERSTOWN AND WINCHESTER.

Via Washington and Metropolitan Branch Road at 6.45 A. M. daily, except Sunday. Via Main Stem at 8 A. M., and 4.10 P. M., except Sunday.

FOR THE VALLEY OF VIRGINIA, VIA METROPOLITAN BRANCH ROAD.

Leave Baltimore at 6.45 A. M., (except Sunday) arriving at Harrisonburg at 4.00 P. M., and Staunton early the same evening.

FOR FREDERICK.

Leave via Main Stem at 8.00 A. M. and 4.20 P. M. daily, except Sunday; and 6.00 P. M. daily.

FOR WASHINGTON.

Leave at 6.50, 8.45 and 11.00 A. M., and 1.00, 3.30, 5.06, 6.30 and 8.45 P. M., stopping at all stations.

Leave at 4 20 A. M., stopping at Relay, Annapolis Junction and Laurel only.

Leave at 5.20, 6.45, 7.40 and 10.00 A. M., and 6.00 and 10.30 P. M., stopping at Relay only.

Leave at 4.00 P. M., stopping at Relay, Annapolis Junction and Beltsville.

Leave at 4.30 P. M., stopping at Relay and Bladensburg.

FROM WASHINGTON.

Leave at 5.00, 6,45 and 8.45 A. M. and 12 M.; 2.00, 3.45, 4.45, 6.45 and 7.45 P. M., stopping at all stations.

Leave at 8.00 and 9.45 A. M., and 1.00, 5.45 and 9.30 P. M., stopping at Relay only.

Leave at 3.30 P. M., stopping at Laurel and Relay only.

SUNDAY TRAINS—WASHINGTON BRANCH.

For Washington—Leave at 4.20, 5.20, 6.45 and 8.45 A. M., and 1.30, 3.30, 5.06, 6.00 and 8.45 P. M., stopping at same stations as during the week.

From Washington—Leave at 6.45 and 8.00 A. M., and 2.00, 4.45, 5.45, 7.45, and 9.30 P. M., stopping at the same stations as during the week, except the 8.00 A. M., which stops at all stations.

FOR PITTSBURG, VIA WASHINGTON AND METROPOLITAN BRANCH ROAD, CONNELLSVILLE ROUTE.

Leave Baltimore at 6.45 A. M. daily, except Sunday, and 6.00 P. M. daily. No change of cars. Pullman cars on night train

Tickets can be purchased at the Office, No. 149 WEST BALTIMORE STREET, corner of Calvert, where orders can be left for baggage to be called for, and which will be checked at persons' residence.

For further information, Tickets of every kind, &c., apply at the Ticket Office, Camden Station.

THOS. R. SHARP, Master of Transportation.

L. M. COLE, General Ticket Agent.

AVAIL YOURSELF

Of the Advantages Offered by the

PITTSBURGH,

Washington & Baltimore

SHORT LINE.

It is the SHORTEST AND QUICKEST ROUTE.

Its Trains are Equipped with

New and Elegant Day Coaches

AND

PULLMAN

Palace Drawing-Room SLEEPING CARS,

Insuring Comfort and Luxury by Day and Night.

TAKE THE

Baltimore & Ohio

AND

PITTSBURGH,

Washington and Baltimore

SHORT LINE.

CONNELLSVILLE ROUTE.

PULLMAN PALACE CARS

AND

Magnificent Day Coaches

FROM

BALTIMORE AND WASHINGTON

TO

PITTSBURGH

WITHOUT CHANGE.

A TRAVELER SAYS:

IT IS THE

UNCONTRADICTED TESTIMONY

OF

TOURISTS FROM ALL LANDS,

THAT THE

SCENERY

OF THE GREAT

BALTIMORE & OHIO,

IN

Natural and Artistic Loveliness,

In all the combined elements of HISTORICAL IN-
TEREST, the BEAUTIFUL, the PICTURESQUE
and the SUBLIME,

STANDS PRE-EMINENT

AND

UNRIVALED

AMONG THE RAILROADS OF AMERICA

And the Traveler knows whereof he speaks.

MONOPOLIZES

THE ONLY LINE THAT RUNS

PULLMAN'S

Palace, Drawing-Room and Sleeping Cars,

AND

ELEGANT DAY COACHES

FROM

Washington and Baltimore

TO

COLUMBUS,

CINCINNATI & ST. LOUIS

WITHOUT CHANGE.

MONOPOLIZES, in fine, all the elements that constitute A FIRST-CLASS RAILWAY.

Secure Your Tickets via the Baltimore & Ohio,

For Sale at all the Principal Ticket Offices.

SIDNEY B. JONES, G. P. A.   THOS. R. SHARP,   L. M. COLE, G. T. A.

Cincinnati, O.   **Master Transp'n.**   **Baltimore, Md.**

THE GREAT
BALTIMORE & OHIO
RAILROAD,

HAVING EVER IN VIEW THE

Comfort and Welfare of its Patrons,

Has already completed, and in course of construction, at convenient points along its line,
some of the

FINEST
HOTELS AND MEAL STATIONS

IN THE COUNTRY,

Where the hungry traveler will be furnished with

Everything the Market Affords,

AND

Ample Time Given to Enjoy a Sumptuous Repast.

Dyspepsia Shops, and the old song, "Fifteen
Minutes, &c.," do not exist upon this Line.

NORTH GERMAN LLOYD.

STEAM BETWEEN BALTIMORE AND BREMEN, VIA SOUTHAMPTON.

The Screw Steamers of the North German Lloyd, KONIG WILHELM I., OHIO, BRAUNSCHWEIG, BALTIMORE, of 2,500 tons and 700-horse power, run regularly between BALTIMORE AND BREMEN, via Southampton.

PRICE OF PASSAGE.—From Baltimore to Bremen, London, Havre and Southampton—Cabin, $100; Steerage, $30. From Bremen to Baltimore—Cabin, $100; Steerage, $40.

Prices of passage payable in gold or its equivalent.

They touch Southampton both going and returning.

These vessels take Freight to London, Hull, Leith, Hamburg, Antwerp, Rotterdam and Amsterdam, for which through bills of lading are signed.

An experienced Surgeon is attached to each vessel.

All letters must pass through the Postoffice. No bills of lading but those of the Company will be signed. Bills of lading will positively not be delivered before goods are cleared at the Custom House.

For freight or passage apply to

A. SCHUMACHER & CO.,
No. 9 South Charles Street.

ALLAN LINE.

STEAM BETWEEN BALTIMORE AND LIVERPOOL, CALLING AT HALIFAX EACH WAY, AND AT NORFOLK, VA., WESTWARD.

The splendid Screw Steamers of the above Line will run every fortnight, taking passengers and freight to and from LIVERPOOL.

PRICES OF PASSAGE:

Baltimore to Liverpool or Queenstown—
Cabin...$75 Gold.
Steerage.. 30 Currency.

Liverpool or Queenstown to Baltimore—
Cabin...$94 50 Gold.
Intermediate... 47 25 "
Steerage.. 32 00 Curr'y.

At which prices parties desiring to send for their friends can obtain tickets.

Through Bills Lading issued to Amsterdam, Rotterdam, Hamburg, London, Antwerp and Havre.

Goods must be cleared at the Custom House before delivery of bills of lading, blanks for which latter will be furnished shippers.

For freight or passage apply to the Agents,
A. SCHUMACHER & CO.,
9 South Charles Street, Baltimore.

THE CARROLLTON,

R. B. COLEMAN, Proprietor,

Baltimore, Light and German Streets,

BALTIMORE.

This new and beautiful Hotel is now open to the Public. It is located on the site on the "Old Fountain Hotel," on Light Street, extended by an elegant front on Baltimore Street, and is convenient alike to the business man and the tourist.

It is the only Hotel in Baltimore of the modern style, embracing Elevators, Suites of Rooms with Baths, and all conveniences; perfect ventilation and light throughout; having been built as a Hotel new from its foundation.

To accommodate Merchants and others who visit Baltimore, the Proprietor will charge $3 per day for the rooms on the fourth and fifth floors, making the difference on account of the elevation. Ordinary transient rates for lower floors $4 per diem.

Guests of the house desiring to avail themselves of the above rates, will please notify the Clerk before rooms are assigned.

An Improved Elevator, for the use of Guests, is running constantly from 6 A. M. to 12 P. M., rendering the upper stories accessible without fatigue.

The undersigned refers to his career of over thirty years as a Hotel Manager, in New York and Baltimore, confident that with a new and modern house, he can give entire satisfaction to his guests. R. B. COLEMAN.

St. Clair Hotel,

GILMOUR & SONS, Proprietors,

Monument Square,

BALTIMORE.

H. H. FOGLE, Manager.

"The Arlington,"

WASHINGTON, D. C.

IS NOW OPEN FOR BUSINESS.

During the Summer, the Hotel has been

Re-Decorated and almost wholly Re-Furnished.

Many changes have been made to add to the comfort of the guests.

The Proprietors are determined it shall not be excelled by any Hotel in America or elsewhere.

T. ROESSLE & SON.

IMPERIAL HOTEL,

WASHINGTON, D. C.

JAMES SYKES, Proprietor.

23

TAYLOR & PRICE,

IMPORTERS AND DEALERS IN

Swiss, French and Nottingham Curtains,

Upholstery Goods, Curtains and Window Shades,

ALL KIND CABINET MAKERS' MATERIALS,

No. 11 N. Charles Street, Baltimore·

MYERS BROTHERS,

Wall Paper, Window Shades, Venetian Blinds, &c.

No. 39 NORTH GAY STREET,

Opposite Odd Fellows' Hall,

BALTIMORE, MD.

B. T. HYNSON & SONS,

MANUFACTURERS AND WHOLESALE DEALERS IN

PAPER HANGINGS, WINDOW SHADES,

MOSQUITO NETS AND WINDOW AWNINGS,

Venetian Blinds, Oil Cloths, &c.

No. 54 N. HOWARD STREET, BALTIMORE.

HUGH SISSON,

STEAM MARBLE WORKS,

COR. NORTH & MONUMENT STS., BALTIMORE, MD.

Marble Monuments, Tombs, Grave Stones, Mantels, Furniture Slabs, Counters, Tile, &c.

COLEMAN & ROGERS,

PHARMACY AND MINERAL WATER DEPOT,

No. 178 West Baltimore Street,

IMPORTERS OF

JOHANN HOFF'S GENUINE EXTRACT OF MALT.

BALTIMORE AND OHIO RAILROAD.

BY JOHN T. KING, M. D.

In April, 1827, the Baltimore and Ohio Railroad Company was completely organized, and Jonathan Knight and Col. Stephen H. Long were selected by the Board of Directors to make the necessary survey of the country through which the road was to be located. The Government of the United States was interested in the great enterprise to such an extent that it detailed several of its chief engineers to aid in the accomplishment of the survey.

In due time the report of these able engineers was presented to the President, Philip E. Thomas, and the Board of Directors, the said report affirming "the entire practicability of a railroad from Baltimore to the Ohio river, along the valley of the Patapsco, Singanore creek, to Point of Rocks in Frederick county."

The construction of the road was commenced on the 4th day of July, 1828, and the event was celebrated with extraordinary excitement and ceremony. The earth was broken and the first stone was laid by the venerable Charles Carroll, of Carrollton, then over ninety years of age, and declared it to be the most important act of his life, surpassing his signing the Declaration, on the southwest confines of the city, where what is now known as Mount Clare, the present site of the immense foundries and machine shops of the Baltimore and Ohio Railroad Company. In August, 1828, the work of grading and masonry was begun between Mount Clare, Baltimore, and Ellicott's Mills, situated on the Patapsco, fourteen miles from the city. On this section of the road, one mile from the city, is the Carrollton Viaduct, a fine structure of dressed granite, with an arch of eighty feet span, over Gwynn's Falls. A short distance further is the famous "deep cut," remarkable for the difficulties it presented in the early history of the road. It is a half mile in length and seventy-six feet in depth. Eight miles from Baltimore you enter the Paloozoic, Plutonian or Granatic region, and in the gorge through which the Patapsco flows the granite formations stand out in bold relief. At this point is the "Thomas Viaduct," a noble granite structure of eight elliptical arches, each of sixty feet chord, spanning the Patapsco at a height of sixty-six feet above the river, and of a total length of seven hundred feet. Upon this bridge or viaduct is the Washington Branch of the Baltimore and Ohio Railroad.

Three miles from the Relay House, on the main stem, is the Patterson Viaduct, a fine granite work of two arches of fifty-five feet, and two of twenty feet span on the river.

At Ellicott's Mills the Frederick turnpike, leaving Baltimore at West Baltimore street, and known as the Catonsville road as far as Catonsville, is crossed by the Baltimore and Ohio Railroad upon the Oliver Viaduct, a handsome stone bridge of three arches of twenty feet span. The road was completed to this point, Ellicott's Mills, and opened for travel on the 24th of May, 1830. In the beginning no one *dreamed* of *steam* upon the road. Horses were to do the work, and even after the line was completed to Frederick, relays of horses trotted the cars from Frederick to Baltimore. At different points along the road relays of horses were provided, and from this circumstance the "Relay House," at the junction of the main stem and Washington Branch, received its name. One great desideratum in the running of the cars, drawn by horses at the rate of eight miles per hour, was to reduce the friction of the axles in their boxes, and this circumstance and difficulty was soon to find a master and remedy. Circumstances undoubtedly make men. About this time appeared in Baltimore Mr. Ross Winans, and with his quick, penetrating, profound and philosophic mind, seized the difficulty by the horns, and instantly became a celebrity by inventing his "friction wheel," an ingenious and beautiful contrivance. The public was intoxicated with the "Winans Friction Car Wheel." The venerable inventor still lives, hale and hearty, and the venerable Charles Carroll of Carrollton, as a "boy again," would take his seat on a little car in one of the upper rooms of the old Exchange building, now the Baltimore post-office and custom-house, and be drawn or hoisted up and down by a weight attached to a string passed over a pulley, and around him would stand, admiringly and delighted, the "prominent and mighty" men of Baltimore, pleased and tickled as children with an amusing toy. Could this same venerable patriot and "signer" of Carrollton leave for a while his sepulchre, and be joined by his old friends, and meet in the "Elevators" of the palace-like structure that bears his name, they would be ecstatic at the luxurious comfort and smooth transportation they would experience in being elevated to the sixth story of the Carrollton Hotel.

When steam made its appearance on the Liverpool and Manchester Railroad it attracted great attention here, but there was a difficulty in running an engine on an American road. The English railroad at that period was made nearly straight, the American road was exceedingly crooked; for a brief season it was believed that this feature of the first American railroads would prevent the use of locomotive engines, but the practicability was soon demonstrated by a gentleman still living at a ripe old age, honored and beloved, and distinguished for his private worth and public benefactions. This gentleman was Mr. Peter Cooper, of New York; he was satisfied that steam engines could be used on the crooked roads already built in the United States, and he came to Baltimore to practically test his faith, upon the Baltimore and Ohio Railroad. Mr. Cooper's engine did not weigh a ton, the boiler was not as large as the kitchen boiler of a range in a modern house; it was about the same diameter, but

2

not more than half as high, "and this was the *first locomotive* for railroad purposes *ever built in America*, and this was *the first transportation* of persons by *steam* that had ever taken place on this side of the Atlantic, and here is the veritable engine, car and passengers, and a perfect likeness of Mr Peter Cooper, standing on his engine, holding the positions of engineer, brakesman and conductor on the first steam train and trip ever enjoyed in America.

This celebrated ride behind Mr. Cooper's engine was between Baltimore and Ellicott's Mills. The open car attached to the engine was filled with the worthy *Directors* of the Baltimore and Ohio Railroad and *their friends*, (this old fashioned example is still in force.) The trip was most interesting, the curves were passed without difficulty at a speed of *fifteen miles an hour*. The *Directors* and *their friends* were elated, and an enthusiast among the Directors, or among *the friends*, when the highest rate of speed was attained, *eighteen miles an hour*, inscribed it in a book to be transmitted to posterity, "*O tempora, O Mores!*" "*Tempora Mutant, et Nos Mutamis cum illis.*"

1873. Baltimore to Washington, 40 miles; time, 39 minutes.

The return trip from Ellicott's Mills was made in fifty-seven minutes, distance thirteen miles, August 28th, 1830.

The first *improved passenger car* built and used in England, and its pattern adopted by the United States, was thus constructed. It was a perfect "*Pullman*" in its day; it was a long box, seats ran along on each side, similar to our city cars, a long deal table was fixed in the centre, and ingress and egress was by a door in the rear, with steps reaching nearly to the ground. The accompanying cut is an exact representation of the first "*improved passenger cars.*"

(SEE NEXT PAGE.)

In 1836 the Baltimore and Ohio Railroad was completed to Harper's Ferry, and the Washington Branch was in operation. The cost up to this time for this distance, Baltimore to Harper's Ferry—eighty-two miles—was $4,000,000.

In 1839 the construction of the Baltimore and Ohio Railroad was commenced between Harper's Ferry and Cumberland, ninety-eight miles, and opened for travel in November, 1842. In 1847 the surveys and construction of the Baltimore and Ohio Railroad were resumed, and the road was completed to Wheeling, June 1st, 1853.

The distance from Baltimore to Wheeling, or the length of the road, is three hundred and seventy-nine miles, and the total cost of construction was $15,639,000. Since the completion of the main stem in 1853, two important extensions have been made by the company, one from Grafton to Parkersburg on the Ohio river, and one from Washington to Point of Rocks, known as the Metropolitan road.

It is the concurrent testimony of tourists from all lands that the scenery of the Baltimore and Ohio Railroad in natural and artistic loveliness, and the sublime, is unrivalled upon this continent or upon the Eastern continent, and for speed, safety and luxurious comfort is not surpassed. The road its entire extent passes through scenery enchanting, wild and sublime; it goes under mountains, around mountains and over mountains, spans rivers and deep gorges by bridges as graceful and airy as a spider's web. For the greater portion of its route it traverses scenery that refines the soul and fascinates the senses; it climbs high mountains, winds along picturesque valleys and keeps company with the crystal Potomac, and looks down upon the silvery "Cheat" and "rippling Tray." Vast mountains scan their dark walls in front, and before one can exclaim "how is it possible to get over that mountain," the train dashes into a dark hole, and instantly total, tangible darkness envelops all, and we dash through the mountain and emerge into daylight on the opposite side, or the engine and train, without halt or hindrance, leap up the mountain side on as perfectly constructed stair steps as those of a dwelling, and descend in the same manner on the opposite

side. Upon one portion of this road between Piedmont and Altamont, you ascend the mountain upon a gradient of one hundred and
seventeen feet to the mile, and this is maintained for seventeen
miles, and for most of the distance the road is constructed immediately over the Savage river, foaming and chafing, seven hundred
feet immediately below the car-wheels; indeed, in some places an
object dropped from the car window would fall clear of the track
and fall clear to the giddy depth below. Between Point of Rocks
and Harper's Ferry, magnificent tunnels are constructed, and the
magnificent bridge and mountain gorge and scenery of Harper's
Ferry are of world-wide notoriety. Between Harper's Ferry and
Cumberland is the great Doe Gulley tunnel, twelve hundred feet in
length, extending under a mountain one thousand seven hundred
feet in altitude. Between Grafton and Parkersburg one dashes
through no less than twenty-three of these long dark tunnels in the
distance of one hundred and four miles, one of these tunnels being
two thousand seven hundred feet in length; but the great tunnel is
the "Kingwood," near the sublime Cheat river region; it is four
thousand one hundred feet in length, cut through a mountain of
solid rock; to make this tunnel it required two years and eight
months of the incessant labor, day and night, of three thousand
miners, masons and laborers. The bridges of the Baltimore and
Ohio Railroad are attractive to the scientist, the engineer and the
picturesque-loving tourist. The trestlings and bridges across the
gorges, especially those spanning the Cheat and Tray, are sublime
in their ethereal position and altitude; they are supported by slender
pillars of cast iron, apparently as light as wire gauze, yet strong and
durable, and almost without a vibration when the ponderous engine
and trains dart across them. One of these bridges is one hundred
and sixty-six feet above the stream and valley, an elevation one foot
more than Washington Monument and its statue;• the other one
hundred and thirty-two feet altitude. One feels in crossing these
bridges as if he was riding in the air, or wheeling amid the clouds,
and is intoxicated by the exhilaration of his sensations, and awed
by the surrounding sublimity.

At Grafton the Baltimore and Ohio Railroad strikes the lovely
"Tygart Valley" river, that for picturesque scenery its entire course
is seldom equalled, and runs parallel with it for nearly one hundred
miles, until it joins the Monongahela; then almost immediately you
come in view of the Ohio, and run parallel with its banks until you
arrive at Wheeling; here the railroad crosses the Ohio upon a magnificent bridge nearly two miles in length, and seventy feet above
the sullen tide of the Ohio. The Baltimore and Ohio Railroad also
crosses the Ohio river at Parkersburg upon a similar bridge.

The hotel system is a feature of this great railroad. The Baltimore and Ohio Railroad Company keep their own hotels; they are
furnished with every convenience, comfort and luxury, and trains
stopping, passengers are allowed ample time to partake of a sump-
tuous meal without confusion, hurry or anxiety. Their "Queen

City Hotel," Cumberland, is a building of magnificent proportions, unsurpassed in every respect; the grounds of the hotel are enclosed and handsomely laid out, and ornamented with fountains, trees and shrubbery, with beautiful lawns and croquet grounds. The centre building is one hundred and forty feet long, two-stories high, surmounted by a cupola; an ornamental piazza ten feet wide extends along the entire front and ends, giving a promenade of four hundred feet; the wings are forty-seven by eighty-four feet, four-stories high; the back-building is thirty-seven by ninety-seven feet, three-stories high, with a basement under the entire building. The entire hotel is heated by steam, provided by two large tubular boilers.

The Company's next grand hotel is the "Deer Park Hotel." This elegant hotel is situated at "Deer Park," Garrett county, on top of the Alleghany Mountains, three thousand feet above the level of the sea. The hotel is surrounded by a magnificent grove of forest trees, and magnificent mountain views are attainable in every direction. The hotel is four-stories high, with Mansard roof, and its entire extent is surrounded by broad piazzas above and below It is elegantly furnished, and is supplied with both gas and water throughout. There are one hundred and fifty rooms, all elegantly furnished. The grounds are tastefully laid off in walks, drives, flower-beds and fountains. Connected with the hotel are livery stables, ten-pin alleys, billiard rooms, and croquet grounds. It is in every respect the peer of the "Queen City" at Cumberland.

At Oakland is the "Glades Hotel," and at Grafton a cleanly and comfortable hotel. With these facilities and comforts, the travel by the Baltimore and Ohio Railroad is truly pleasurable and luxurious. At all these hotels the mountain air is bracing and life-restoring, and to the over-worked, invalid and tourist, they furnish a delightful retreat, where the refined and intelligent meet in social intercourse and enjoy the magnificence and sublimity that environs them on every side.

The Baltimore and Ohio Railroad, ninety miles west of Grafton, passes through Moundsville near Wheeling, and the tourist comes in view of the curious and ancient tumuli or mounds, from which structures the town derives its name. They were built by an ancient and extinct race, of whom tradition nor history furnish any information. That they were an intelligent and artistic people, ample testimony proves. These mound builders must have been numerous, and probably possessed and occupied the vast area between the Alleghany and Rocky Mountain regions, for their mounds and various articles of their handicraft are to be found over a great portion of the western area of the continent. They existed long previous to the creation or advent of the North American Indians, and far exceeded them in intelligence and artistic skill. These mounds are about seventy or eighty feet high, and the same in diameter at base. They are nearly perfect cones, some of them being truncated by time and the attritive agency of rain. They are hollow in the interior, having only one chamber; and in these chambers have been found the osseous remains or portions of the human skeleton. Sometimes these bones are of gigantic size, and indicating their living possessors to have been a race far greater in stature than any present existing, or historically described people. Along with these human remains are occasionally found coins, perfect in their workmanship and finish, bearing visibly characters and inscriptions that closely resemble, if they truly are not, the ancient Runic Nurnigraphy of the extinct and warlike Norsemen, that one finds upon quaint old rune-stones or coins. Beyond a doubt, whoever these mound builders were, they lived long previous to the Indian race, and were versed in abstruse science and some of the fine arts; and it is to be regretted that no reliable tradition or written record preserves their name and deeds. And a similar fate awaits their successors—the Indian race; in a few years, within half a century, not one will be left; no vestige will be found either in the way of architecture, art or written record of their own, to furnish evidence or remind the nations of the earth that a mighty race had at one time existed, and possessed and occupied the vast American continent. It is sensible to pause and consider, Two great nations once on this continent totally extinct, buried beneath the deep tide of oblivion as unknown, and as little considered as the dead beasts of the field, or the hidden carcasses of the sea. But what matters it to them? human remembrance and human praise or condemnation affects them not; they were the creation of their Maker; they, in their existence, subserved his purpose, and that accomplished, the mound builders and the Indians glide out of existence as the fading scenes in " dissolving views " of the camera.

VALLEY RAILROAD.

ROBERT GARRETT, PRESIDENT.

The Baltimore and Ohio Railroad, joined at Point of Rocks in Frederick county, Maryland, by its Metropolitan or Washington branch, pursues its course to Harper's Ferry. At Harper's Ferry it enters the Shenandoah Valley by its Winchester, Potomac and Strasburg connection and uses this road as far as Strasburg, fifty-one miles up the Shenandoah Valley. At Strasburg the Baltimore and Ohio transportation is continued upon the Manassas division of the Washington, Virginia-Midland and Great Southern Railroad, and this connection preserves the line of the Baltimore and Ohio Railroad, unbroken, as far up the valley as Harrisonburg, in Rockingham county, a distance of one hundred miles. Heretofore the Baltimore and Ohio Railroad, by contract only, made use of the Manassas division of the Washington, Virginia-Midland and Great Southern Railroad, for the purpose of transportation between Strasburg and Harrisonburg, a distance of fifty-five miles; but under a recent lease the Baltimore and Ohio will have sole control and assume exclusive jurisdiction over this division. The Valley Railroad connecting Harrisonburg and Staunton, thereby effects uninterrupted transportation from Staunton to Baltimore and Washington city by the new route. Staunton is located on the Chesapeake and Ohio Railroad, which connects Richmond and Huntington, West Virginia, by a line four hundred and fifty miles in length, extending entirely across the States of Virginia and West Virginia, and traversing every variety of country and through mineral sections of great wealth. From Staunton, it is known, the Valley Railroad is to go south, the work being now under contract, and will traverse Rockbridge county to Lexington, pass through Botetourt county, and terminate at Salem, in Roanoke county. The entire length of the Valley Railroad from Harrisonburg to Salem is one hundred and thirteen miles.

At Salem the Valley Railroad connects with the Great Atlantic, Mississippi and Ohio Railroad, reaching southward to connections with the Virginia and Georgia Railroad and the East Tennessee and North Carolina Railroad at their junctions in the State of Tennessee.

The Valley Railroad in its entire extent traverses the Valley of the Shenandoah, a region celebrated for its fertility and agricultural wealth, its picturesque valley and magnificent mountain scenery. From Harper's Ferry to Salem the valley of the Shenandoah is enclosed on the east by the long blue South Mountain range, and on the west by the North Mountain chain, both being ridges of the great Blue Ridge or Appalachian range. The whole valley, averaging twenty-five miles in width, is highly cultivated, luxuriant crops of cereals and hay occupy the soil, and numerous fat herds of beef cattle are to be seen browsing and feeding in broad, luxuriant pastures.

CONSTRUCTION AND MASONRY OF THE VALLEY RAILROAD.

From Harrisonburg to Staunton, twenty-six miles, the Valley Railroad is graded, ballasted with the exception of a few hundred yards, and substantially bridged. There is not upon the continent a road more thoroughly and substantially constructed; the road bed is enclosed by two parallel walls formed of large stones and stone slabs, similar to the curbs of a street, and within this space is deposited the comminuted limestone fifteen inches in depth; upon this substantial bed the crossties are placed, and between the crossties the spaces are filled with the same material, thus forming a solid, undisturbable foundation. Along the entire line of the road are inexhaustible quarries of the finest quality of limestone, and the blasting, cutting and excavations through them were a stupendous and laborious undertaking. Many of the quarries along the road furnish a variegated limestone, beautiful in its variegated laminæ and striæ, and of sufficient strength and density to be utilized in the way of slabs for tables and furniture, and the construction of mantles, and the fissilliferous formations exhumed in its construction would enthusiastically entertain and engage the geologist.

The direction of the Valley Railroad takes it across turbulent rivers and wide chasms, and causes it to pierce mountainous bluffs and obstructions. For variety, picturesqueness and sublimity it is not surpassed by the Baltimore and Ohio save in its Cheat river scenery. It traverses in its entire extent a region of the valley unsurpassed in rural loveliness and landscape beauty. Valley and mountain are blended in one view. On the one hand the intensely blue South Mountain trends away towards the south in graceful undulations, and the "Three Sisters" affectionately nestle side by side, and the dark black wall of the "Massinutton" towers aloft unto the clouds, and when gilded by the rays of the setting sun presents a scene of splendor that dazzles by its gorgeousness and awes by its sublimity In its course the Valley Railroad going north from Staunton crosses the Middle river, and crosses the North river three times, necessitated on account of the tortuosity of the stream. Hence it was necessary to construct four large bridges to span these rivers. In the construction of these bridges the masonry of the Valley Railroad is certainly unsurpassed. In the execution of the work masons who have been engaged on the Mount Cenis tunnel were employed. The skill of the artisan and quality of material excite and attract admiration, and the labor, science, and engineering skill displayed upon these gigantic bridges will ever be a monument to the engineers of the Valley Railroad. The Middle river bridge is 450 feet long, 60 feet above the water, and has two abutments and three piers. The three North river bridges are respectively 315, 340, and 350 feet long, 40, 50, and 61 feet in height, and have two piers and abutments each. The deep, lengthy cuts upon some portions of the road are wonderful. At section 11 the road is cut through solid rock and limestone 60 feet high, and this cut is over 600 feet long, and the cut and road gracefully conform with the curve in the river, which is immediately beneath the track. At Mount Sidney, ten miles north of Staunton, is the magnificent cut a half mile in length and twenty-four feet deep, through the solid rock, and penetrating a bed of limestone rich in fossiliffeous treasures. The culverts are also of the most substantial masonry, and every portion of the road bed and construction presents a thoroughness and durability seldom found. The cost alone of these bridges was from $50,000 to $60,000 apiece, and the average cost of construction of this portion of the Valley Railroad was about $35,000 per mile, and when one considers the amount of work done and the obstructions surmounted, the construction cost is surprisingly economical, and is an evidence of the judicious expenditure of the funds and the faithful discharge of the respective duties of all concerned.

At Harrisonburg, the eastern terminus of the Valley Railroad, the road bed is constructed through the western suburbs of the town, and by its construction that portion of the town has been very decidedly improved. Streets have been straightened and graded to conform to the level of the road bed, and substantial stone bridges built across muddy depressions, and culverts of substantial masonry have been constructed by which thorough drainage is secured. In the portion of Virginia and West Virginia through which this road will pass, are 21,000 square miles of the richest mineral deposits upon the earth, coal, iron ore, magnetic iron ore, hematite, emery, and a region inexhaustible in the finest timber for the ship builder, carpenter, or cabinet-maker. The coal fields of this region exceed by double the amount of square miles those of the State of Pennsylvania, and will for centuries furnish a home and employment to thousands of colliers and miners. The valley will furnish the mining emigrant from every portion of Europe an abundant and healthy home, and the mountains of coal and iron and timber will guarantee work and remuneration for untold generations yet to come. In these mountains at present game is abundant, and the virgin fertility of the soil in the valleys will furnish readily all that is required for the sustenance of man. There is not upon the earth in the same given space such a variety and combination, such abundance of mineral wealth, or a spot more charmingly inviting for plentiful and happy homes. It may truly become the paradise of the miner and collier, and in due time this wild region will swarm with the stout laborers of Europe and our own country.

2*

With the contributing lines of communication at Salem the Valley
Railroad, when finished to that point, will carry the products of the
south and southwest and receive at Staunton the contributions by
the Chesapeake and Ohio Railroad, and at Harrisonburg the inexhaustible products of the vast coal and mining regions of West
Virginia by the narrow gauge, and from thence transport them
over the line controlled and worked by the Baltimore and Ohio
down the valley of the Shenandoah to Harper's Ferry, and thence
by the Baltimore and Ohio Railroad to the depots and wharves of
that company. The advantages of the great consuming and dis-
tributing capacities of Baltimore will thus be realized to great and
distinct sections by this combination of routes.

MARYLAND.

BY JOHN T. KING, M. D.

If one will contemplate the physical formation of this continent,
it will be immediately and strikingly observable that there are two
great water sheds, one on the Eastern or Atlantic littoral margin,
the other on the Pacific or Western confines of the continent; and
that the continent is traversed by two vast mountain ranges, main-
taining very nearly, throughout their entire extent, a northeasterly
and southwesterly direction, very nearly equidistant, and including
a vast valley, whose transverse area is the distance between the
Alleghany and Rocky Mountain ranges.

In this area, from the Atlantic to the Pacific, are to be found
every phase and formation, geologically, from the lower Silurian
up through every variety of formation and strata to the most recent
or alluvial.

In the locality and within the limits of Maryland, the whole
geological series is present, and after leaving the ancient primitive
formations and Silurian system on the Atlantic water-sheds, one
enters the cretaceous and carboniferous area in the Alleghany
Mountain range. East of Dan's range of the Alleghanies, as far as
the Blue Ridge, all is cretaceous, confined to the mesozoic period;
west of it, the formations belong to an earlier period, the paleozoic
or carboniferous era, and it is here in Maryland that one finds
himself into the midst of the great coal fields.

That this region was, at one time, submerged, that even the
loftiest peaks of the Alleghany Mountains were for a considerable
time covered by the sea, is indisputable. The fossil remains of
both existing and extinct marine molluses, univalvular and bival-

vular and piscatorial fossils are daily met with in the carboniferous formations, firmly and deeply imprisoned, and extricated by the miner's pick.

That the great upheaval of this mountain range was long subsequent to the coal formations, requires no argument or extraneous proof to convince one of the fact. Upon the apices of the loftiest peaks of the range are to be found the coal measures or beds, with their strata undulating conformably with the irregularities of the range throughout its longitudinal concatenation, gracefully curving perpendicularly in some places, in others, breaking off abruptly, the terminal escarpments of the fracture being several hundred feet above and below a horizontal line, one of the escarpments being on top of the mountain, the other from several hundreds to several thousands of feet in the valley.

That they have been submerged but *once*, there is ample proof, and undoubtedly this long submergence was very early in the mundane existence. It was away back in the silurian and paleozoic ages, when the earth was in its infancy, and it extended throughout the entire carboniferous era. Of this submergence and upheaval, no tradition could ever exist, or within record testify, for the human family was not in existence. The Deluges, traditional and scriptural, are occurrences of yesterday, compared with the ages and eons of time that have elapsed since the Alleghany Mountains were covered by the sea. The human family did not come into existence until the long mesozoic period had elapsed, and more than two-thirds of the cenozoic had passed away: not until the almost termination of this latter period, amid the eocene, miocene and pliocene ages, did the human form appear on the earth.

That the valleys of the Alleghanies have been inundated by pluvial and fluviatile accumulations, there can be also no doubt. The irruption of bodies of water through the mountain ranges is apparent, and admits of logical proof. In a number of localities, contiguous to one another, bodies of water contained in the deep valleys between the mountain ranges, have worn a channel through the mountain and these vast lakes have been drained and the ancient water-beds are now dry land and fertile valleys. Between Piedmont and George's Creek the Savage River, or natural water flow or drain of the eastern aspect of the Glade's region, worked its way, conjointly with the Potomac, through the transverse sections of Dan's range, and the pent-up waters of the vast lake, now valley, between Dan's Mountain and the great Savage Range, and between Dan's Mountain and Will's Mountain, worked through, by solvent and altritive agency, Will's Mountain, one mile from Cumberland, leaving a gorge indescribably sublime and known as the "Narrows." Innumerable instances of the coercive outlets through the mountain ranges exist in Maryland, Pennsylvania, and Virginia, the escaping waters leaving the ancient Lacustrine bed dry and arable land. That these ancient lakes were filled with water up to the level of the lowest depression in the mountain range, and that by the attrition or wear of their overflow the gorge in the mountain proceeded from *above* downwards, until a bed or level was attained as that of the lake-bed, is also apparent. The gorges are all wedge-shaped, or infundibuliform, the base of the wedge being uppermost, or as an inverted cone.

If the irruption of the waters of these lakes had begun *subterraneously*, the disintegration of the mountain, and the enlargement from the attrition of the escaping waters of the escapement channel would have been almost exclusively lateral or horizontal, for the weight of a volume of water flowing through a channel acts more effectually upon the bed and sides of the passage through which it flows, and as the lake would soon be lessened in volume by the outflowing of its water, the flow through the forced aqueduct would be lessened and subsiding from the roof of the tunnel would leave an arched-way perfect and continuous through the mountain. The formation and existence of cascades and large water-falls prove conclusively that the attrition of overflow and irruption commences at the *summit* of every gorge. The rapids or escapement of waters of Lake Erie is corroboration of this fact. The eastern and western hemispheres are vast, bold mountain ranges, and the bed of the Atlantic Ocean is a vast valley three thousand miles broad by ten thousand long. The Atlantic Ocean is a vast lake or series of lakes, precisely similar in its topography to Lake Erie and Ontario, for there is a great water-fall or Niagara, a great precipice extending from the eastern to the western hemispheres, from Newfoundland to Ireland, known as the great telegraphic plateau, on account of the Atlantic cable being laid upon it. This Oceanic or inter-continental precipice is overwhelmingly sublime, not only on account of its great horizontal extent, but also as to its fathomless depth, the depth of the ocean, instantly and as abruptly as the Niagara Precipice, increases from a few hundred fathoms to a depth of seven miles, and soundings there are not obtainable. The beholder is mute viewing the Horse-shoe Falls at Niagara; the effect would certainly be insupportable, if he could stand on the floor of the ocean immediately beneath this oceanic cataract, and see a sheet of water three thousand miles broad falling from a perpendicular height of probably ten miles or more.

There is no question but what the vast valley between the Alleghany and Rocky Mountain ranges was at one time an inland sea almost equaling in extent the great valleys or waterbeds of the Atlantic and Pacific Oceans. From the incessantly changing character of nature, the molecular restlessness of matter, constantly altering in form and position, this interchange of land and water that once has taken place, will undoubtedly occur again: what is now dry land will be submerged and the Atlantic and Pacific basins be dry and arable valleys and broad prairies dressed in gorgeous floral garniture; where now these profound waters roll, and the Leviathan and whale and smaller fishes disport and have a home, broad crops of golden sheen will wave like the undulations of the sea, and resplendent cities will arise, and happy homes will exist, and millions of the human family and animals of every type and species will live and move.

The mountain ranges now covered by the sea will stand forth in grand and lofty ranges with their fossil stores of fish and shell and animal remains. The Azores, Teneriffe, and Mederia, will be snow-clad peaks piercing the hovering clouds ; the Pacific archipelagoes will be one grand mountain range broader in area and greater in altitude than the Alleghanies or Rocky Mountains. On this ocean bed will iron bands be laid upon which lightning trains will dart. Man will wander amid this wreck and ruin of a pre-existing continent and race and curiously and enthusiastically explore the long avenues, deep chambers and broad arches of the coral mountains that will dazzle the eye by their purity and whiteness as they gleam in the splendor of the noonday sun, and as he traces and gazes on this interminable line of coral pinnacles and spires that trend for thousands of miles, along this coast, they will appear like the ornately chiseled nave of some vast and gorgeous cathedral, and he will be alternately fascinated and awed by its beauty, magnitude and grandeur. He will doubtingly pause and ask, could this have been built and adorned by the insect world ? Man will also roam amid groves and interminable forests of gigantic ferns and trees, and numberless herds will browse and feed upon the, at the present time, subundine luxuriant foliage and herbage, that detached and withered, has for ages, and is now adrift, appearing like a mid-ocean prairie in the verdant Sargasso sea, deceiving Columbus by its land-like appearance nearly four centuries ago, when traversing an unknown sea, by inspiration, making for an unknown world.

On the other hand, the races that inhabit this now dry land will be no more! their cities will be submerged, and fish will disport in their chambers and streets, and on the broad plains and in the long valleys in place of the buffalo, there will roam countless and huge monsters of the deep ! where now the husbandman sows and reaps his crops of grain, countless acres of molluses and fish organisms will exist. The Alleghany and Rocky Mountain ranges will be islands of the sea, uplifting their peaks above the circumfluent waste of waters.

Lastly—when this new continent appears, formed of the Atlantic and Pacific Basins, the Scientist, the Naturalist and the Geologist of that period fraternally and inquisitively roaming, will find the exuviæ shells and skeletons of once living and curious forms and species, and they will confer and reason and concur, determine and affirm that these now submarine mountains and valleys were once covered by the sea, nor will they desire or demand any human tradition or written record to aid them in their conclusions or to prove the fact, for the position and enduring testimony will be inscribed upon the mountain sides and coral reefs, and an imperishable record will be graven in the fossiliferous pavement of a sea-abandoned continent.

By what agency or force this alteration of sea and land will be effected, it is reasonable to conjecture. As the former interchange was occasioned by violent, igneous, volcanic and gradually exerted calorific force, so, in all probability, will causes again effect a like result. That it will again occur is undoubted, whether the event shall be remote or imminent, for to the watchful observer and geologist, the gradual interchange has already begun and gradually and surely going on : encroachments of the sea upon some coasts in some localities is yearly apparent and has attracted attention from the thoughtful and observant, and promising and proving a verification of the Scriptures that this earth shall not wholly be destroyed again by a universal cataclysm, but by another agent—a consuming, pyrogenous one.

The climatology of Maryland, as influenced and determined by its position and physical aspect, is of necessity varied, including a range thermometrically and hydrometrically from arctic to equatorial conditions. The eastern shore and the eastern belt that littorally extend along the Chesapeake Bay, are exceedingly humid and temperate, almost tropical in its seasons and flora. The eastern shore, sensibly feeling the thermal influence of its inter-oceanic position and proximity to the Gulf Stream, the winters are more mild and humid and vernal season or weather is several weeks ahead of other portions of the State ; the northern and western portions are cold and rainy, and long and vigorous winters prevail. In the Alleghany region the springs are late and chill, and rain descends without warning, owing to some drifting clouds getting in the cold condensing stratum of the mountain range.

The prevailing winds nearly all over the State are almost diametrically opposite, previous to, and succeeding the winter and summer solstices, on the eastern shore. The winds are much influenced by and indeed are partially the northern limits or caudal fragments of the great Atlantic "Trades" that unchangeably blow from the same quarter, the southeast, and absorbing heat in its passage across the Gulf Stream, averaging two hundred miles in width, infringes upon our coast laden with warmth and moisture. Under these circumstances, this eastern shore region must of necessity be a locality highly favorable to agriculture, horticulture and fruit culture, independently of any natural fertility or fertilizing agent. These climatic advantages, combined with the deliciousness, variety, and abundance of its marine and fluviatile productions, render th s portion of the State the Paradise of the epicure and gentleman of ease, and the Mecca of the literary and lazy.

CHAPTER II.

That portion of Maryland intervening between the Chesapeake Bay and the Blue Ridge range of mountains, especially that portion north of the Patuxent River as far as the Pennsylvania line, belongs to the ancient lower and upper silurian paleozoic period. Indeed, no portion of the earth in the Eastern or Western hemisphere can claim seniority to it, in this area, especially about the City of Baltimore and Ellicott's City. All along what is known as Elk Ridge are formations exclusively restricted to the lower and upper silurian age, and the fauna consist totally of the extinct fossil specimens of molluses and fishes. The region referred to is upon the eastern or Chesapeake Bay's littoral confines, from the Patuxent river to the Patapsco, bluffy, especially in the counties of Prince George's and Calvert, and so marked in this physical characteristic, that the southern portion of Calvert is called the "Cliffs of the Patuxent."

Each of these counties is jurassic and cretaceous of the ovlitic and mesozoic period. This elevation is maintained as you recede from the bay shore, and is diversified by undulations, increasing in magnitude, until you strike the Blue Ridge or Appalachian range proper. This region, in a metalliferous point of view, contains some iron ore, and near the City of Frederick are to be found slate quarries of a superior quality of that mineral. Going directly west from Baltimore, the first glimpse of the Blue Ridge range of mountains is obtained at Frederick city. Here a long, lofty ridge rises into view, trending north-east and south-west, and is known as the Catoctin Mountains. Beyond this is another range which is the "South Mountain." Crossing this range you descend into one of the loveliest valleys that mortal eye ever rests upon, containing fine residences and highly cultivated farms, and presenting every landscape and rural beauty, that nature and cultivation can bestow. It was in this valley, sheltered on the east by the South Mountain and on the west by the North Mountain, that one of the most sanguinary conflicts of the late fratricidal war took place. The immediate locality of the battle was adjacent to the little village of Sharpsburg, and upon the banks of the picturesque Antietam creek. Upon the site of this deadly struggle is the National Cemetery, and contains inhumed within its precincts the remains of over five thousand warriors. In this valley also is situated the neat and urban like town of Hagerstown.

Along the summit of South Mountain is presented a most curious and interesting phenomenon, and which is readily resolved into the certainty of its being the bed of an ancient river. It runs parallel with the mountain range and its longitudinal inclination or dip is toward the north, consequently the waters of this ancient river must have flowed in that direction. Apart from this northern inclination or dip, the disposition of the stones, boulders, and debris, prove conclusively that such must have been its direction. The boulders and stones lie in an implicated position, like the alternate courses of shingles or ties upon the roof of a building, and some of them show the corrugations of a tide or current ripple upon their surfaces, undoubtedly impressed thereon when in a plastic state, presenting the familiar appearance, such as one may see at any time, upon the sandbars and sand shores of our rivers and sea coast. Undoubtedly, the bed of this ancient river was, in its entire extent, elevated to its present position on the summit of South Mountain, when the elevation of the mountain took place, and presents all the appearances of a dried up stream.

Beyond Frederick city as far as Harper's Ferry, the Potomac river flows through a channel interposed between the north and south escarpments of the Catoctin and South Mountains. At Harper's Ferry the South Mountain abruptly terminates, or is apparently cleaved in twain. The mountainous height on the Maryland side of the Potomac is known as the Maryland Heights, and towers aloft in an almost perpendicular escarpment. Upon this apparently inaccessible height were encamped the armies of the North and South alternately, and the heaviest ordinance was transferred to its summit. That part of the mountain on the Virginia side of the Potomac is known as Loudon Heights, and is the resumption of the range which extends southerly into Virginia. It is at this point, Harper's Ferry, that the Potomac and Shenandoah effect their confluence, and apparently by irresistible and combined force have rent the vast mountain in twain, but the impression that either of those rivers forcibly burst through the mountain is without reasonable foundation, and can be controverted and proven by the simplest reasoning. The Shenandoah flows along a natural bed, or a longitudinal depression, the valley of Virginia, and this depression or river bed is continued at Harper's Ferry in the gorge between the Catoctin and South Mountains. This depression or natural bed is not interrupted at all at Harper's Ferry by any mountainous obstacle, nor ever has been. The river would flow, if it had never united with the Potomac, naturally around the base of the abrupt Catoctin Mountain. The Potomac also flows in an original and natural bed, and at Harper's Ferry follows the depression between the north and south extremities of the South Mountain. If the two rivers, or either of them, had by irruptive force pierced the mountain, large quantities of immense boulders would have been precipitated into the bed of the streams and translated shorter or longer distances by the impetuous current; but such is not the case. The sides of the mountain forming the walls of this gorge are almost perpendicular, and the bed of the river is paved with slabs of sandstone, which originally existed in the stratum. Moreover, the edges and angles of the rocks, that constitute the wall of the escarpments, show no evidence of being smoothed or rounded by the attrition of water or any other agency, nor have the channels of either of the rivers widened by cavings-in or wear, within the recollection of man, and when annually the most impetuous and irresistible torrents rush through these gorges.

As before stated, geologically, Cecil, Baltimore. Harford, portions of Anne Arundel, Howard, Montgomery, and Frederick counties belong to the several different periods of silurian, permian, metamorphic, and jurassic. There have been found, and can always be found the most ancient formations and types of molluses, fishes, reptiles, birds, and quadruped animals, whose existence stretched over a space from the paleozoic through the mesozoic to the end of the cenozoic era. The Blue Ridge Mountains constitute an abrupt and perfect line of demarkation between the ancient silurian and paleozoic formation and period, and, west of a line drawn from Harper's Ferry to the Pennsylvania border, is a totally different geological formation. On the east of this line you leave the ancient formations referred to; on the west you immediately enter the permian and calcareous eras, the upper and lower chalks, and find yourself in the midst of the mesozoic age. The formations in Washington and Alleghany counties are almost totally limestone and sandstone, outcropping in strata in every exposed situation. After crossing the intermediate valley or plain between the Blue Ridge and the Alleghanies, you enter at once into the mountain region of Maryland, the North Mountain forming the eastern confines of the great Alleghany range. Following the Potomac, when one arrives at Hancock, you instantly enter the mountain fastnesses, and from Hancock to Grafton, or nearly a hundred and fifty miles, the traveler is in a maze and labyrinth of mountain peaks and ranges, in the labyrinthian gorges and windings between the mountains and the Potomac river. The Baltimore and Ohio Railroad and the Chesapeake and Ohio Canal are running parallel and side by side.

The city of Cumberland is situated in a basin or concavity completely surrounded by mountain peaks and ranges. On the next side of the city the lofty and magnificent Wills Mountain rears its dark green wall; to the south, the lovely Nobleys erect their serrated forms, and away beyond, surmounting all and like a monarch among mountains, rises the sublime and towering Dan's Mountain. One mile west of Cumberland, Will's Mountain is completely transversely divided by Will's creek, and through this sublime gorge passes the creek, the National road and the Pittsburg and Connellsville Railroad, all running parallel and side by side, and conforming to the graceful curvatures of the gorge. The transverse section of the mountain is one mile in length by about three hundred feet broad, and in this gorge or the "Narrows," as it is called, is one of the most interesting and sublime pieces of scenery that is to be met with on this continent Upon either side are the truncated extremities of the mountain, nine hundred feet in height, with almost perfect perpendicular escarpment, and the summit presenting a castellated appearance, that it is difficult to realize are not genuine castles. The capping and strata of this gorge is a reddish sandstone, and the great altitude and fantastic forms of these strata cause the illusion, at one point, to be perfect. There is the turreted castle, jutting out from the dizzy cliff with bastions and columns, and one can easily imagine himself in the presence of some fierce warrior and in the domain of some lordly knight. From Cumberland, westwardly, one can traverse one of the loveliest valleys that human vision ever beheld. It is a valley lying between the Nobley Mountains and Will's and Dan's Mountains. It is about one mile wide and near thirty in length;

through it flows the Potomac river, and parallel with the river runs the Baltimore and Ohio Railroad. The mountains on either side are high to dizziness and gracefully serpentine in their range.

At New Creek and at Piedmont, and next to Piedmont as far as Altamont, the scenery is overwhelmingly sublime, and it must be a callous soul that would not be attuned to solemn and adoring mood. It is at Piedmont, the long upward seventeen miles grade of the Baltimore and Ohio Railroad commences, and the ascent or gradient of the road is one hundred and seventeen feet to the mile. On this grade, on the ascending journey, two locomotives have to be attached to the train.

On the right of this grade is the great Savage range of mountains, and three hundred feet immediately below the railroad track is the Savage river. From the height whence one views it, it has the appearance of a silvery band, winding at the base of and among the fastnesses of the mountains.

It is at this point, in this awfully sublime and overwhelming scenery, may be seen the lonely residence of ex-Governor Francis Thomas, of Maryland, at present United States Minister to Peru. He fled to these mountain fastnesses in the depth of his soul's grief, with crushed heart and hopes, that his broken spirit might receive consolation in the lonely communion with nature in her sublimest aspect. Anchorite, monk or cynic could not desire nor find a locality more sublime or lonely. He was prompted to this solitary abode on account of his divorce from his young, sprightly, and beautiful wife, whose fascinations of mind and person charmed and graced Maryland's and Virginia's most aristocratic and intellectual circles. He loved his lovely bride with an insane ardor ; but the demon could not let the doting statesman enjoy his earthly happiness. Jealousy, discord, and finally separation, closed the scene. She married, soon after, a Presbyterian clergyman, and has since resided in Philadelphia.

At Altamont, 3,100 feet above the sea level, you enter upon a wide level plain known as the glades. This glade region is the summit of the Alleghany region. It is near Altamont the interesting phenomenon is seen of two streams or rivers running in opposite directions; running east is the Potomac, and the Youghioghany within a few yards running west. The Potomac courses to the Chesapeake Bay, in which it empties its waters at Point Lookout, and the Youghioghany flows towards and discharges into the Ohio. The head waters of the Potomac have their source a short distance from Altamont, near Fairfax stone, a stone that indicates the boundary between Virginia and Maryland.

Reclining by the tiny and sparkling stream and drinking from the crystal rills that confluently make up the grand and lengthy Potomac, I paused to reflect and felt pleasure in contemplation and realization of the fact, that I had, in five weeks time and tramping, traversed over five hundred miles, and had kept company with the noble river from Point Lookout to the tiny rivulet at my feet, and I felt an affectionate emotion as I laved my brow in the fountain from which my lengthy companion took his source. In turning from and bidding adieu to my companion for so many days, I felt saddened, feeling that I had heard for the last time its musical and soothing murmur, and should no more wander along its lovely banks. And you, dear old Virginia, the mother of States and of statesmen, what a sisterly feeling should exist between you and Maryland, claiming in common the noble Potomac, across whose waters these sublime mountains have held for ages solemn and silent converse, and upon whose common soil the defenders of your homes and firesides unsheathed their swords and crimsoned the soil of each with the gore of your noblest sons. As a wall of fire, your sons stood to stem the infuriated tide of a powerful and relentless foe, and your noble mountains and once peaceful valleys glared in torchlight of the incendiary, shrieked at the wail of the widow, and wept at the plaint of the orphan. People of a common origin and blood, with your sister Maryland, descendants of the lordly pilgrims who unfurled to the breeze the banner of liberty and knelt at the foot of the cross at Jamestown and St. Inigoes, I must bid adieu to your hospitable mountain homes and the scenes amid which I so love to linger. Their impression and remembrance will gladden my pathway through life, and cause my glazed eye to brighten with a gleam of joy in my extremity.

CHAPTER III.

It is in Alleghany County, the extreme western county of Maryland, that you enter the intensely interesting regions of that great geological era in the wonderful and sublime cosmogony. It is here you are first introduced into the great coal measures, and brought in contact with the ancient carboniferous formations.

As an abrupt and almost perfect line of demarkation exists, formed by the North Mountain, between the watershed of the Blue Ridge Mountains, extending eastwardly to the sea, and the plain or valley interposed between the North Mountain and the Alleghany range on the west, so, geologically, the area east of Dan's Mountain or range is totally different from that west of it. On the east of Dan's Mountain, all the geological formations and strata are limestone and sandstone, geologically, comparatively recent. On the west is to be found a more ancient formation, viz: the great coal measures.

The great coal basin lies chiefly between Dan's Mountain and the great Savage range, occupying an area transversely of about five miles on an average and about sixty in length. In this valley or basin between these two mountain ranges flows a tortuous and shallow stream, known as George's Creek. In periods of drought nothing is to be seen but a dry, rugged water bed, but in seasons of rain and melting snows this channel is flooded by a deep and turbulent torrent. The sources of George's Creek are near Frostburg, and after traversing this valley for seventeen miles, anastomoses with the Potomac at Piedmont.

On either side of this George's Creek valley, its entire length, the mountain ranges are filled with coal to their very apices, and the coal measures or beds underlying the valley and beds of the creek to the depth of one hundred and sixty-five feet. Shafts are sunk into the earth to the depth of nearly two hundred feet, through which the coal is elevated to the surface. Where the mine is on top of the mountain the coal is transported down the mountain side in small cars regulated by a steam engine, on a tram-way, laid on the mountain side.

There are innumerable mines in this George's Creek valley, and from Mount Savage to Piedmont is one continuous street and town, twenty-four miles in length, inhabited by miners and their families. The miners are almost all Welsh and Scotch.

Having had my curiosity gratified in the interior of the earth at a depth of one hundred and sixty-five feet, my aspirations took the opposite direction—to the mountain top. At the invitation of the gentlemanly superintendent of one of the mines, I concluded to accomplish the feat, the ascent of Hampshire Mountain.

I was introduced to the obliging boss of Hampshire Mines, and was by him directed to take a standing position in one of a train of small cars. At a signal to the engineer, stationed on top of the mountain, off we started. We glided up and adhered to the side of the mountain until we reached the first level or halting place, five hundred feet from the surface of the earth. Landing on this level, we were transferred to a small locomotive and proceeded horizontally around the mountain, when we arrived at the starting point of another perpendicular ascent. I mounted to one of a train of small cars, similar to the one in which I had made the first stage of the ascent, and off we started on the upward journey of two thousand feet, almost at a perpendicular. In this, the second stage of the ascent, I became dizzy, blind, and nauseated, and when I arrived at the top of the mountain I felt as relaxed and looked as exsanguined as if I had been seriously ill. And the view from this mountain top, who can describe? Its equal has never been painted on canvas. On every hand I was encompassed by an illimitable sea of mountains, and the long deep valleys appeared like fathomless sinuses or troughs between the mountain billows.

At the top of the mountain, at the entrance of the shaft, the second boss of the mines courteously inquired : "Are you ready, sir, to go into the mines? the guide is ready." Answering affirmatively, the boss delivered minute instructions to the guide to remain as long as I desired and to conduct me into every portion of the mines. My cicerone having received his instructions, off we started, each carrying a torch, elevated above our heads. We proceeded along the main avenues, but frequently turned off to explore some branch avenue and chamber. These tunnels through the mines are about fifteen feet wide, by eight or ten in height, and are cut through the solid coal. The floor of each tunnel is laid with an iron railway, upon which the small coal cars run. The cars are drawn by horses, and each car and horse has an attendant or driver. To the cap of the driver and to the side of the head of the horse is attached a lamp, which is necessary, as the darkness is unutterably black and tangible. At one point, according to the estimation of the guide, we were fifteen hundred feet under the mountain, a grave deep enough, and a superincumbent weight sufficient to confine and silence, one would think, the mightiest and most refactory demon.

In the centre of this mining region is situated the town of Frostburg. It is fourteen hundred feet above Cumberland, eleven miles distant, and two thousand three hundred feet above Baltimore. The town is completely underminded, all the coal having been removed from beneath it, with the exception of the columns of coal left as supporting pillars to uphold the terrestrial shell and town. The Cumberland and Pennsylvania Railroad passes under the town in a subterranean tunnel. In going from Cumberland to Frostburg, eleven miles distant, the whole way is up mountain, and to accomplish the ascent the road is constructed in the form of Y's; the locomotive and train climb one Y, switch off on another, and so on, backwards and forwards, until at last you arrive safely on the top of the mountain and at Frostburg.

A view unsurpassed for panoramic character and grandeur, save one, is obtainable two miles from Frostburg, from the summit of the great Savage Mountain, from which point of observation, one can look upon a sea of mountains rolling in Pennsylvannia, Maryland and Virginia, and in the town of Frostburg, one thousand feet below, I could count every house. What pen can describe or pencil portray the grandeur and beauty of this mountain region? It is God's unrolled canvas, upon which, with a Master's hand, He has tinted and touched the whole into transcendent loveliness and sublimity. But not alone to please the eye of man has He fashioned these mountains and painted them in emerald, purple and gold, and decked the valleys with flowers rich in fragrance and gorgeous in bloom. He has made these mountains the storehouse for the useful and indispensable coal that blazes on every hearth-stone, that warms and gladdens alike the rich man's palace and the humble cottage home, that furnishes the gas light to illumine man's pathway and make resplendent the domicils and halls of the wealthy and refined. Prescient and merciful God! What is man that thou shouldst be so tenderly mindful of him? that thou shouldst make the mountains and valleys and the fathomless sea and the vast aerial ocean all subservient to his use and pleasure? Thoughtless and ungrateful man! canst thou not comprehend that all these things are of God's infinite goodness and love?

Traversing this mountain region, independently of the Baltimore and Ohio Railroad, are two important and noteworthy highways, ancient routes coeval with the republic, one of them, the other antedating it and contemporaneous with the regal sovereignty that exercises its power over the infant American Colonies. One is Braddock's road, that extends from the District of Columbia to Fort Du Quesne near the present city of Pittsburg on the Ohio River. This road was constructed by Braddock, assisted by the youthful Washington, for the purpose of transportation of troops to overcome the French forces occupying the country of the Ohio and Monongahela rivers, and to suppress the depredations of the hostile Indian tribes intervening and existing in that locality. Every inch of the route was over almost, one would think, insuperable mountains, through an interminable wilderness through the domain of hostile Indian tribes, across morasses, jungles, and rivers for a distance of over four hundred miles.

The undertaking and construction of this road would, one would affirm, intimidate the bravest heart and paralyze the stoutest arm, and it is astonishing how cognizable this road is in some portions of its route, when one considers that nearly one hundred and twenty years have intervened since its construction. Through the courtesy of Dr. Charles Getzendanner, of Frostburg, as cicerone, I was enabled to traverse a portion of this road, south of and adjacent to Frostburg. On the side of the road stands emplanted a dark gray slab or tablet, about two feet wide by three in height. On the reverse of this tablet is inserted in old English characters, " 11 miles to Fort Cumberland. 29 miles to Capt. Smyth's Inn and Bridge— Big crossings. The best road to Red Sandstone. Old Fort, 64 miles." On the obverse is inscribed in the same character of letters, "Our Country's Rights we will defend." Either the iconoclast or sacred memento-loving stranger has chipped off the angles and edges of the venerable tablet. There can be no reasonable doubt that Washington saw and touched the ancient landmark, and who can say that he did not, with his own hands, erect it, and chisel the letters, plainly visible upon its weather-beaten and gray surface ?

A short distance west of this, on Laurel Mountain, Braddock received his mortal wound. Disregarding the cautious admonition of his young, but sage Lieutenant, George Washington, he attacked the savages in their ambush and fell by the unerring arrow of the Indian. Braddock reposes in death's embrace in a little valley in the noble Alleghanies, and Washington in the hearts of his countrymen, in the soil of his beloved Mount Vernon.

Along this road made by Braddock and nearly equidistant are vestiges of forts, constructed of stone, and some of them in a wonderful state of preservation. There is one near the town of Hancock that astonished me by its size and state of preservation. The next important fort was erected on the heights in south Cumberland, overlooking Wills Creek, and upon its site is now erected the Episcopal Church; also upon this site, ancient Fort Cumberland, are the vestiges of a well-constructed fort by Braddock and Washington, although the sparkling waters of Will's Creek laved the base of the fortress of the ancient Fort Cumberland. The garrison were compelled to abandon its use on account of the unerring and deadly Indian arrows, that were showered upon them whenever they emerged from behind their ramparts, except in full armor and force.

The other interesting trans-Alleghany highway is the old "National road." This great road in days of yore had its eastern terminus at the General Wayne Tavern, northwest corner of Baltimore and Paca streets. This road left Baltimore in the route of the present Baltimore street, and at the western limits of the city was known, and is now known as far as the city of Frederick, as the Frederick road or Turnpike. Westward of Frederick it has maintained its baptismal name of National road. This road extends from Baltimore over the Blue Ridge and Alleghany Mountains to Indianapolis, Ohio, to the distance of nearly five hundred miles, and was, previous to the Baltimore and Ohio Railroad, the great highway and only thoroughfare between the east and far distant west.

"Leaves have their time to fall," and the old National road has sunk into oblivion and disuse—has had to yield to the iron band that encircles almost the globe. A short time ago and the last stage coach, the last of its memorable race, rumbled over its time and travel-worn pavement, and when I saw it draw up in front of the post office in Cumberland for the last time—the end of its last journey from Frostburg, I lingered around it with reverence and affection. Here it was, the very type of the venerable and ancient diligence, low swung by broad leathern springs on low cumbersome wheels, corpulent in body, with a great capacious boot in front and the mail apartment in the rear, covered by a dusty, crispy leathern apron, fastened by large rusty buckles and broad rusty straps : such was the appearance of the last of the old great stage coaches of the great National highway; and when it was driven into the old stage yard for the last time and its wheels were locked never to revolve again, a feeling came over me akin to the melancholy solemnity of the obsequies of an old departed friend, and I turned away in sadness. And how they will be missed along that ancient highway, in the villages and towns, where their arrival and departure from the doorway of the post-house tavern with its big, swinging sign and ruddy, burly landlord, was an event exciting and momentous; and in these old post-taverns, in those old by-gone days, what news was circulated and startling stories related by the social travelers seated before the wide extended fire-place, and cheered and comforted by the flaming, snapping, roaring, great log fire.

It was my happiness to sit in the porch of one of these old stage taverns in the suburbs of the town of Frostburg, on top of the great Savage Mountain, and listen to the exciting narrative of an old stage driver upon the great National Road, recounting with rapt pleasure the hair-breadth escapes and deadly encounters with the mail robbers and highwaymen of this, then, wild region. It was blissful to sit in the full moon's soft glow, flooding with a silvery light the sweet and peaceful valley below me, and illumining the long lofty range of the great Savage Mountain beyond, and with solar brightness defining the dizzy peaks of the great towering Dan's Mountain at my side, turbaned by the gauze-like cloud that in fleecy texture floats above his majestic head, and like the nuptial veil of some fair bride, gracefully sweeps like gossamer adown the great mountain's side, and so poetic, peaceful and sublime the scene! The full-orbed moon poised in the zenith all aglow, the giant mountains all around and the shimmering moon-lit valleys stretching far away. No noise disturbs the tranquility of night. My friendly dog would coil himself upon the grass for sleep, and nought would be heard save the old stage driver's cracked, tremulous voice, and the tinkling of the bells in the sheepfold in the valley.

After crossing the great carboniferous belt, which, previous to the division of Alleghany county, was principally confined to that area, but now is included by both Alleghany and Garrett counties, you traverse, in going due west, a number of mountain ranges, increasing in altitude until you arrive at Laurel Ridge or Laurel Mountain, some thirty miles west of Frostburg and the great Savage range. These mountain ranges on the north extend into Pennsylvania, and on the south into Virginia, and after crossing Laurel Ridge, decrease in altitude, until they become the foot-hills of the Alleghanies that margin the Ohio River, and at Wheeling forming an eastern mural inclosure to that city. But the foot-hills of the Alleghanies are not abruptly estopped by the Ohio River, for on the Ohio side of the river undulations and elevations extend far into the interior of the State, like the exhausted undulations or ground swell of a tempestuous sea. The geological formation of this region, west of the coal measures or the great Savage range, is now again resumed or identical with those east of the Dan's Mountain.

You again, immediately after crossing the Savage range, enter the cretaceous era and mesozoic period, and have almost abruptly and totally left behind or east of you the Paleozoic and carboniferous age and formations. The mountains are almost totally limestone and sandstone, covered by a dense growth of oak, maple, white pine and other forest trees, and the laurel undergrowth occupies the mountain sides to their very apices in impenetrable density, and furnishes a perennial food for the deer and pheasants that abound in this region; and the caverns and interstices among the rocks, boulders, shale, and debris furnish an impregnable abode to the rattlesnakes and copperheads that thrive and swarm in this region. A few bears, also, and panthers, reside in or frequent these fastnesses, and for miles these animals and reptiles are the monarchs and occupants of this wild mountain region.

About seven miles west of Frostburg, on a plateau upon the summit of the great Savage range, and skirting the old National road, is a locality called the "Shades of Death." For several miles the old road was darkened by the dense growth and deep gloom of a white pine forest, and the entrance to this realm would cause the heart of the old National road traveler to palpitate with fright and his voice become husky. And no voice was heard during the transit of this sepulchral portion of the old National road, but that of the old stage driver, urging and cheering his nervous team, with cocked pistol in his belt and eyes right and left, for here in the "Shades of Death" was the favorite rendezvous of the old highwaymen and mail robbers of those good old times. For one, I shall never forgive the authors, corporators, and capitalists, who had nothing else to occupy their idle brains about, than project and build the Baltimore and Ohio Railroad. This great iron road has been the death blow to all the poetry, romance, and enjoyment of travel, and at this day not one in a million knows of or has ever taken that grandest of journeys in this wide world, from Baltimore to Wheeling, across the Alleghanies, or passed through the "Shades of Death."

From the city of Frederick, whose eye ever beheld a landscape more glorious than that "Middletown Valley," and crossing the South Mountain, and looking upon the Hagerstown or Cumberland Valley, and the gore-stained field of Antietam, one beholds a picture that is unspeakably lovely and enchanting; further westward, at Clear Spring, on the south escarpment of the North Mountain, where on earth can such a scene be found? One mile from Cumberland you pass through a gorge in Will's Mountain, the "Narrows," that cannot be surpassed in the Rocky Mountains or any other region; then you climb to the summit of Dan's Mountain, and here let us pause at Frostburg, and in every direction from Frostburg scenery gorgeous and sublime greets the vision. And the "Rock" on the summit of Dan's Mountain—who has touched? Not a score of Marylanders or others.

By arrangement the previous evening, my friend and guide, T. W. Clary, of Frostburg, and I, arose at 4 A. M. and mounted our horses for the journey to the "Rock" on the summit of Dan's Mountain, the distance up the mountain from Frostburg being seven miles. After leaving the town we took the Piedmont Road, and passed the numerous coal fields and the great "Bordan Shaft," 160 feet deep, through which the miners descend into the mine, and up which the coal is brought to the surface in small cars hoisted by a powerful steam engine. Turning to the left at the "Shaft," we immediately began the ascent of Dan's Mountain. From the tortuosity and roughness of the mountain road, it being obstructed the greater portion of the distance by boulders, rocky fragments, and general debris, we made slow progress, as it was physically tiresome to the horses. Apart, in common with ourselves, they felt the effect of atmospheric attenuation, which is peculiarly exhausting. Sitting in our saddles, we were panting and gasping, and had to breathe about forty times per minute, instead of the normal amount of respiratory effort, or twenty per minute, so as to get the necessary amount of oxygen in our blood.

In our ascent, the first noteworthy locality we came to was the village of "Pompey Smash," the home of several hundred miners

3

and their families, Welsh, Irish and Scotch. The Pompey Smashers
are a gay and festive population, devoted to "red eye," "mountain
due" and "morning star," and to form a just estimate of Pompey
Smash character and society, one must see them at either a wedding
or a wake, or their monthly free fight. On either of these occasions
they may be seen in all their glory. Apart from their hilariousness,
they are an honest, hard-working village of miners, and each pater-
familias possesses a wife, a three hundred pound porker, a game
rooster, and from nine to fourteen children. Pay-day, the fifteenth
of each month, is the day that dawns brightest upon them, and on
that morning the festivities of Pompey Smash begin, and if a Sunday should succeed within a day or two of the fifteenth, they are
ecstatic. The corn juice, Scotch reel and Irish jig absorb every
Pompey Smasher, old and young, and the grand climax and finale
of the festivities is the ecstatic "free fight," which causes Dan's
Mountain to leap for joy. These monthly jubilees result in a trans-
mogrification physiognomically ; the Scotch visage and accent are
made broader, the Irish features are flattened, and the Welsh nasal
appendage is softened down several degrees, and his voice !—every-
body knows what the pronunciation and accent of Welsh is like—
you would be sore in body and brain for a month to hear a Welsh-
man talk five minutes. In these social "Germans," the ladies of
Pompey Smash are not neglected. They participate in the flowing
bowl and terpsichorean evolutions *sans* clogs or stockings, and en-
liven the occasion with the loud, mirthful blending of Welsh, Celtic
and Scandinavian vociferations, with bare arms and shillelah accom-
paniment.

After leaving Pompey Smash, we climbed along a road wooded
on both sides with chestnut timber and oak, and after a three hours
ride, came in sight of the lonely, gray, jagged and towering rock.
We tethered our horses at its base, climbed from ledge to ledge up
its rugged side some fifty or sixty feet, and reached the level summit.
What description can convey the slightest idea of the illimitable expanse and sublimity all around ? I was overwhelmed with emotion
and sank down upon the rock in silence and awe, perched amid the
clouds, whose gauze-like texture floated about my head in a bound-
less sea of mountains, the beautiful Piedmont valley four thousand
feet below, and the Potomac like a silvery ribbon winding through
it, and the Baltimore and Ohio Railroad, and trains, the cars of
which did not look larger than a train of black ants. We drank in
the cool, pure air of the heavens, and gazed on the river-like appearance of the moving mist, as it stretched away and curved and rolled
around the mountains, and settled like a fleecy, wide-spread table-
cloth on the surface of the deep down valley. We sat in silence and
contemplated the scene, filled with the blended impression of pleasure, solemnity and awe. On this "Rock," "which like a giant
stands to sentinel enchanted land," should arise a fane whose spires
and domes should be enwreathed by and pierce the fleece-like clouds.
It is a site fit for the throne of a monarch whose sceptre should sway

from the frozen regions of the north to the Patagonian shores, and from the waters of the Pacific to the Atlantic's western confines.

In Garrett county, the extreme western county of Maryland, called so in honor of John W. Garrett, Esq., President of the Baltimore and Ohio Railroad, and also in commemoration of a deceased brother, Henry W. Garrett, whose beneficence erected a substantial and ornately constructed Gothic church in the village of Oakland, the same geological series are present, and impressively noticeable are the perfectly cubic-shaped and immense blocks of limestone, scattered over that region. They are in the valleys and on the loftiest mountain peaks. The regularity of their figure and angles are remarkable; a stone-mason could not have chiseled them more geometrically regular. Some of them, thousands of tons of weight, are poised on the very top of the mountain, and look as if they could be tilted off and sent crashing down the mountain side by the power of a child's finger.

Twenty-two miles west of Oakland, the mountain rover or the tourist in Pullman Palace car, whistling through the tunnels and mountain gorges, halts at Rowlesburg, the eastern introduction or environs of the unequalled "Cheat River" scenery and region, that will alternately fascinate, awe and affright, for here Nature simultaneously writhed in agony and split her sides with laughter, and touched with her finest pencil the gorgeous scene.

CHAPTER IV.

There is not in this wide world a valley so sweet
As that vale in whose bosom the Cheat and Tray meet.

At Rowlesburg, on the Baltimore and Ohio Railroad, 253 miles from Baltimore city, the tourist in the elegant coaches of that greatest of railroads on this round world, or the foot-sore and sun-bronzed geological wanderer and searcher, laden with rocky fragments in pockets and hat crown, enters the portal of the sublime and ravishingly sweet valley where the Cheat River and Tray Run effect their confluence. The topography east of Rowlesburg and immediately around the village, although mountainous, foreshadows nothing of the grandeur and sublimity that suddenly unfold to view a few hundred yards west of the village; this grandeur and sublimity are compressed and crowded into the space of seven and one quarter miles, from Rowlesburg to Tunnelton, and for that distance there is not, cannot be found in the wide world, the sublimity, grandeur, and beauty, so impressively and harmoniously blended. The dark, steel-colored "Cheat" gently flowing through the narrow valley hardly one hundred yards wide, and the silvery Tray rippling through the mountain gorge to mingle with the Cheat; the sublime mountains, swelling perpendicularly from the side of the Cheat, and lifting their domes and spires three thousand feet am d the clouds, the ever varying scene, at morning, noon, and evening,

by sunlight or moonlight, and the flitting cloud, will instantly
cause a totally different aspect and effect on river, mountain, and
valley. I have witnessed the mellow, soft, deep twilight, almost
amounting to darkness, instantly settle over this little valley and
stream and the vast mountain, caused by the flitting, interposed
cloud, and there is not the same aspect and effect maintained for
five minutes at a time in this treasure of the Alleghanies, and gem
of the Baltimore and Ohio Railroad. In viewing the Cheat River
from the tresseling of those two bridges over the Cheat and Tray,
the mind is rapidly and alternately oscillating between awe,
sublimity and the impression of ravishing loveliness, and one is
fascinated and charmed, and the senses are intoxicated amid this
profusion and feast of nature ; and one wishes to linger yet awhile,
another day, to stay there, to die there, in this deep mountain
gorge, by the side of the murmuring Tray. But it is at night,
when no sound is heard in this deep, lone valley, save the
ripple of the cool, silvery Tray, when the round, full moon peeps
over the high wall of yonder big mountain, and her rays fall aslant
up the little valley that the fairy-like scene becomes ecstatically
lovely. It is at this period that the pencil or pen is utterly im-
potent, and the mind is lost, intoxicated, abandoned, and drinks in
the scenic draft to delicious unconsciousness, and is only aroused
by the shrill scream of the "fast line," climbing up the mountain
side, hanging half way up the mountain on a ledge five and a half
feet wide. What! a train of passenger coaches and Pullman cars
and a magnificent locomotive climbing a mountain side on a road
bed five and a half feet wide and seven hundred feet above a yawning chasm! Travel yourself over this road that has not its equal
on earth, in equipment, management, and scenery, and I will
venture to predict that no investment of the same amount of $22.00
can be made, that will yield an interest so substantial, instructive
and enjoyable. Its beneficial and pleasurable effect will abide with
you to life's close. At this time, the truly humane, benevolent and
beneficent of our city are carrying out the noble and praiseworthy
undertaking of "free, indigent children and mothers' excursions,"
whereby they can be enlivened and revivified by the pure air of
God and delighted and cheered by the sun's rays and the beautiful
blue heavens, that they never saw or heard of before, in the narrow,
noisome, pestilential alleys in which they were born and live, amid
squalor, crime and woe. This act upon the part of these noble
women and men of our city, I verily believe, will endear them to
God infinitely more, and cause His heart to expand with
mercy toward them, and His countenance to shine with more
benignancy than all the college and hospital buildings that could
be constructed over a space ten miles square. Next to this let a
grand excursion be inaugurated to extend from Baltimore to the
Ohio River, on the Baltimore and Ohio Railroad, and let the in-
vitations include every poor person who has never been over the
road, men and women, and children over ten years, those that can

appreciate, and give such persons an opportunity to get a whiff of pure mountain air, view the sublime mountain scenery of the Alleghanies and have an idea what the Baltimore and Ohio Railroad is. I believe it would be incalculably beneficial, physically and instructively to the children of our Public Schools to travel over this road. Let each class, annually, go over and return. It would be refreshing and exhilarating, they would enjoy the ride in the easy, fine coaches of the company, and be charmed and excited by the ever varying magnificent mountain and valley scenery of the route.

It has been a matter of astonishment to me that the Baltimore and Ohio Railroad Company, among other sites for their unequaled railroad hotel palaces, as one will find at Cumberland—"The Queen City," and the elegant "Deer Park," and "Grafton Hotel," have not fixed upon one amid the Cheat River scenery, as a location for one of their great railroad hotels. There is the wildest, sublimest scenery on the road; game abounds all the year round, deer and pheasant hunting; fishing all the summer; and a point nearly equidistant between Baltimore and Cincinnati, here, would certainly meet the eastern and western bound tourist, and they would sojourn in scenery sublime and enchanting, in deep solitude and shade, and pure, cool mountain air, that causes one, almost without exception, any night, to sit by a blaze of faggots on the hearthstone, and to sleep with blanket and coverlid. It is here in this valley of the Cheat, the toil-worn and care-worn could have a respite, the drooping be revived, and the contemplative and thoughtful have surroundings of beauty and grandeur upon which their minds would dwell and feast, and be attuned to adoring mood, and directed and uplifted to God by His awe-inspiring works. At present there are no accommodations for the traveler to the valley of the Cheat, except at one or two inferior hotels or boarding houses at Rowlesburg and Tunnelton. The location for a hotel would properly be about midway the valley, or near where the Cheat and Tray meet; there the mountains swell up almost perpendicularly from the water, the whole length of the valley, and in the centre of the valley one is completely circumvented by grand and lofty mountains, so close together and precipitous as to almost exclude the sun, save at noonday. At Tunnelton begins that sublime piece of masonry that is truly awe-inspiring in its length, construction and the human engineering skill, perseverance and labor requisite to overcome nature. Here nature seemed determined to arrest the further progress of the Baltimore and Ohio Railroad; she, directly in the teeth of the great road, billed up or rolled a vast mountain, apparently in desperation, determined that the great work should there stop, and man should succumb to her obstacles and insuperable barrier. She yielded to the Baltimore and Ohio Railroad for over one hundred and fifty miles; her rivers had been spanned by magnificent iron bridges, some of them swinging at dizzy heights from mountain side to opposite mountain side, across deep chasms and turbulent torrents. The railroad had been built around her mountains in a spiral route; her mountains were pierced, bored and tunnelled; she had yielded to the invincible engineer; had been tortured, pushed aside, ignored and laughed at in every conceivable way; but here at Cheat River, the engineer and his army of 3,000 men, with their crow-bars, picks, shovels, and engines, halted, dismayed, disconcerted, and apparently overcome at last. They looked wistfully, silently and inquisitively in each other's faces; a council of war was held; in that council it was determined to "take that mountain." A charge was ordered, and with uplifted ax, shovel, pick and spade, on that army rushed, and the bowels of that mountain were laid open, and the work of evisceration did not cease until the rays of light from the east and west met in the narrow, one mile in length, "Kingwood Tunnel." Here nature gave up the ghost, dismayed, discomfited, and overwhelmed; from there to the Ohio River she ceased to annoy or place obstacles of any extent in the path of the Baltimore and Ohio Railroad.

The "Kingwood Tunnel" required 3,000 laborers constantly employed for the space of three years. Its length under the mountain, from the eastern to the western orifice, is one mile, and its masonry can not be surpassed for safety and durability. Surpassing the Kingwood Tunnel in engineering skill and human perseverance and conquest over almost insuperable obstacles, are the tresseling and bridges over the Cheat River and Tray Run. Description fails to convey an appreciable and acceptable picture of these dizzy and sublime pieces of masonry, and those iron structures; one must be suspended almost in the clouds in a railroad car, to appreciate the great work. You can not see the track upon which the train glides; you look up and see the sky and the dark mountain side across the Cheat; you look down from the car window and you become giddy; the Cheat and Tray are away down below you, immediately under the track, and the tops of the loftiest trees, away down underneath in the valley, look like the surface of cut velvet, and as evenly clipped. The Cheat and Tray have a steellike lustre, and look like coronal metallic bands encircling the base of the dark green mountain. There is not such a grouping of nature's sublimity and wildness, and man's genius, skill and engineering power, over apparently insuperable obstacles and barriers, to be found on the whole earth, within the space of seven and a quarter miles, as is here presented in this Cheat River valley and Kingwood Tunnel. Spanning these rivers with bridges resting upon tresseling one hundred and seventy feet high, looking like wire gauze when viewed a short distance off, blasting a roadbed in the solid rock half way up the mountain side, the outer ends of the railroad ties overhanging an abyss seven hundred feet deep, was victory over nature overwhelmingly sublime. It was a gigantic contest for the right of way, but the engineer came off victorious, and the mighty engine, with its long train, darts up that mountain's side, regardless of the yawning chasm beneath, and spurns the mountain in its pathway. God-like, truly, was the mind and omnipotent the power that overcame the obstacles in that Cheat River valley. The tresseling and tunneling in that locality will be a monument through all time to the genius and skill of those civil engineers who accomplished the work.

I have been whirled around that mountain, seven hundred feet in the air, and dashed across that tresseling, apparently as fragile as a spider's web, and through the labyrinth of mountains, with travelers as unimpressionable as a dead negro, seemingly as insensible to the sublimity that surrounded them, and as unconcerned as if passing over the Eastern Shore Railroad from Crisfield to Delmar, or through the Pine Barrens along the Weldon Railroad in North Carolina. Some people are never excited or aroused to an appreciation or comprehension of the beautiful or sublime, or calculate and consider the herculean, physical and brain forces that were involved and expended in any great work. I have often, in surveying a car of passengers, come to the conclusion that the material composing their bodies would subserve a more useful purpose if it had taken the direction of Cincinnati pork or sauer kraut. I can not stomach such *compagnons du voyage*; it is nauseous and repulsive to be compelled to tolerate the coarse, senseless jest and harsh, vacant braying of a swaggering, garrulous crowd. It is on this account I detest the village, town, and city, and long for the simplicity, quietude and sublimity of nature's great temple. I can only worship in a perfectly adoring mood at the altar of the great I AM, and only feel in befitting place when I am curiously wandering through nature's vast mausoleum, examining the debris and ruins of time, deciphering characters impressed and graven in the imperishable rock, and translating the earth's cosmology written upon the mountain side. It is amid such ruins and records, and such scenery as that about Cheat River, so masterly touched by the Almighty's hand, here in the gloomy shadows of these vast mountains, in that deep gorge by the side of the sparkling Cheat or rippling Tray, that I would love to dwell; here I could uninterruptedly commune with nature, translate her great book and keep my soul unsullied by non-intercourse with the world.

From the mouth of the Susquehanna river and source of the Chesapeake Bay, projects a long narrow peninsula, two hundred miles in extent, by an average breadth of about thirty miles, known as the Eastern Shore. The southern extremity of this peninsula forms the two Eastern Shore counties of Virginia, Accomac and Northampton, the latter terminating in the projection known as Cape Charles. This peninsula, apart from the two counties of Virginia, contains nine counties of Maryland, and is washed on the eastern side, its entire extent, by the Atlantic Ocean.

Distant from the main land of from five to twelve miles, is a chain of islands, extending from the Capes of the Delaware to the mouth of the Chesapeake; some large, others of small area, and separated by inlets. On the western confines of the peninsula is the Chesapeake

Bay, averaging in width about twelve miles, and two hundred miles long, this Peninsula—the Eastern Shores of Maryland and Virginia —is intersected, on an average of ten miles of space, by rivers and creeks, flowing into the Chesapeake Bay and transversely crossing the peninsula to within a few miles of the Ocean on the east. There is no area of the same extent probably upon the globe that is possessed of such maritime and fluviatile advantages and can present in its continued length the same littoral line or surface. From the mouth of the Chesapeake Bay to the mouth of the Susquehanna, the rivers and creeks abound in the finest species of wild fowl, their sub-undine surfaces are paved with the most highly flavored and delicious bivalves, and their waters teem with many species of delicious fish.

This "Eastern Shore," from being interposed between the Chesapeake Bay on the one hand and the Atlantic Ocean on the other, it can be readily perceived that its climate must be mild and appreciably temperate. Lying not more than two hundred miles distant from the Gulf Stream, that courses parallel with its entire extent, it perceptibly feels the thermal influence of that warm oceanic artery. Owing to this inter-oceanic position and consequent mollification of its climate, it is readily acknowledged to be a most favorable locality for the growth of all grains, fruits, and crops. Especially is its benignancy manifested in the early and rapid (compared with other regions) germination, fructification, and maturation of fruits and berries of every variety and species, and it must, ere long, be a region devoted to the table-vegetable productions and become one vast luxuriant market-garden.

Geologically considered, the Eastern Shore of Maryland and Virginia is of very recent formation, and is undoubtedly formed by the silt and detritus swept down by the floods of the Susquehanna and strewn along for two hundred miles, gradually accumulating in depth and breadth, until it finally appeared above the level of the sea. The soil is exclusively alluvial, formed entirely of sand, clay, and humus, and the only organic remains ever found embedded, are the exuviæ of present existing species and varieties of shell-fish and marine and fluviatile vertebrata. Its era is confined to the posteriary period or formations, almost recent enough to be within the memory of a few generations past. Superficially it is level; slight undulations are met with, but no elevations pertaining to hills or mountains. The oceanic littoral line is beveled gradually, and the mariner will find soundings of from five to ten fathoms twenty miles distant from the land. No remains of inorganic substances of either granitic, carboniferous or any paleozoic specimen has ever been found, nor is there any indication of its existence at the time of the earliest flora and fauna.

The earliest possessors and inhabitants of the peninsula were various tribes of Indians, a few of them powerful and numerous. The most powerful tribe occupying the southern portion of the peninsula, was the Nanticokes, and evidences of their occupation are still extant in the oyster-shell mounds and embankments which they left behind, and frequently contain specimens of their trinkets, pipes, and arrow-heads. They must have been almost exclusively piscivorous, as fish were the only article of food bountifully supplied and easily obtained. Up the peninsula other large tribes dwelt, and chief among them were the Delawares, who left not long ago and emigrated to the northwest. These Indians of the Eastern Shore gave names to the various localities in which they lived or frequented, and the Anglo-Saxon owner of the soil to-day, calls and knows them by their aboriginal nomenclature. To the majestic bay that bounds it on the west they gave the name of Chesapeake—the great salt bay; Chingoteague indicates where pike fish are plentiful and are caught; Annemessex, the creek where logs are obtained for building; Choptank, the river with the big bend, and every one who has ever been up that river will recognize the applicability of the name, this big sweeping bend occurring about two miles above the town of Cambridge; Monie, the place of assembling for their big talk on important and state occasions; Nanticoke, the first or big tribe; Pocomoke, the river abounding in shell-fish; Quantico, the big dancing place, where they assembled for the great annual dance; Sinepuxent, filled with oyster beds; Susquehanna, the river with rapids; Tuckaho, where deer are scarce or difficult to obtain; Wicomico, where wigwams or Indian houses were built, an Indian city; Manokin, the place for scalping and where also they had a fort or place for defence; Mytipquin, the great burying-ground where all the dead were buried, the great Indian cemetery. In confirmation of this fact, the author of this article, in visiting the locality, which is a narrow isthmus between the Nanticoke and Wicomico rivers in Somerset county, saw exhumed bones and fragments of human crania, undoubtedly of Indian type. No more convincing, conclusive proof of the existence of an extinct race exists than the osteological remains embedded in the earth. This locality, apart from being their big or great burial-ground, must also have been thickly populated or frequented by the Nanticokes and perhaps visiting tribes; for nowhere on the peninsula, that I have visited, are found more oyster shells embedded in the earth, both in riparian situations and also remote from the water. These Eastern Shore tribes were locally nomadic; they roved up and down the peninsula and from their fierce and warlike propensities frequently came in deadly collision. It sometimes occurred that tribes crossed the Chesapeake Bay, from the Western Shore of Maryland and Virginia, and made incursions into the country of the Acconiacs and Nanticokes, either from war-loving or predatory propensities.

Upon one of a chain of small islands situated in the Chesapeake Bay, about midway between the mouth of the Pocomoke and Potomac rivers, are evidences of Indian occupation, either for resident or warlike purposes. There is a tradition entertained at this day among the inhabitants of this island, that the two powerful tribes, the Powhattans of the Western Shore, and the Nanticokes of the Eastern

Shore, met here in deadly conflict, which battle forever cooled the warlike ardor and humbled the prowess of the Powhattans, and left the Nanticokes master of the field and their domains secure from vexatious incursions and ruinous spoliations. These various tribes in the course of time were mingled into or joined the warlike and powerful Delawares, and after this consolidation the Susquehannas joined their forces with them, and being encroached upon and annoyed by the presence of the white race, they simultaneously struck their wigwams and started for new and far distant hunting-grounds, towards the setting sun. And their race is almost run; a few more full moons will come and go and the descendants of the Nanticokes, the Delawares and Susquehannas will be no more. They will only be remembered as the hideous Indian with his dress of skins, his painted visage, his bow and arrow, his revengeful and bloodthirsty nature; *his* wrongs and the *Christian* white man's deception, treachery and cruelty, will not be remembered nor mentioned.

It is on the northern shore of the Sassafras river, which river divides the counties of Kent and Cecil lying on the north, that we are brought in contact with ancient formations, formations dating back to the metamorphic and jurassic periods in the earth's cosmogony; and one will only have to ascend the Susquehanna a few miles, as far up as Havre-de-Grace and Port Deposit, when he will be confronted with the most ancient works and objects in existence. Here one will see the granitic formations, and be carried back in his contemplations to the earliest period of time. Here is the dividing line between the ancient formations and the new, recent, or alluvial deposits. This line of geological demarkation runs nearly due east and west from the neighborhood of Philadelphia to the city of Baltimore, all north of it being ancient, that south of it being recent or newly formed. On the Western Shore of Maryland this line of separation deflects southerly, and this deflection is maintained entirely across Maryland and Virginia. Calvert county and the counties south of it along the shores of the Chesapeake are recent, or tertiary, and nearly geologically of the same constituents as those on the opposite side of the Chesapeake Bay. Physically the land on the Western Shore is more elevated and undulating, and can be accounted for easily, as being the foot-hills of the Blue Ridge or Appalachian range, from forty to fifty miles distant.

In Baltimore county the formations are of the most ancient type, and along the streams flowing in that and the adjacent counties, westwardly, evidences of the fact present themselves in the exposed strata, granitic outcroppings and boulders that one meets with in every direction. As on the Eastern Shore, large and powerful Indian tribes dwelt upon and roamed over these grounds: so also on the Western Shore there is not a creek, river, or bay, that does not furnish evidence of their possession or occupation. Upon the shores of the St. Mary's river, in the county of St. Mary's, was first unfurled the banner of religious liberty, and where the first English Catholic pilgrims made their humble genuflections at the foot of the cross. At this spot are visible the mural vestiges of the first governor's house of the Lords Proprietary, Leonard and Cecilius Calvert, the Lords Baltimore. Fortuitous circumstances caused it to be abandoned, and the present site of Leonardtown chosen instead.

West of the bay side range of counties, one leaves the recent or tertiary formations and enters upon a broad expanse of ancient structure, and the realms of the earliest types of animal and vegetable organizations, substances, forms and structures, that date their origin and bear testimony to their existence at a period briefly subsequent, when the earth was without form and void, and darkness spread over the face of the deep. They point to a period long away, when no man existed, or any animal or plant; they gloomily, silently and sadly bear record that ages long and countless have elapsed, an ocean of time has intervened that science and the mind of man cannot encompass.

HISTORIC HOMES IN THE VALLEY OF THE SHENANDOAH.

The tourist upon the valley branch of the Baltimore and Ohio Railroad, leaving Harper's Ferry can alight at Winchester or Summit Point and there get conveyance, and a ride of twelve miles in view of the magnificent ranges of mountains will bring him to the most interesting historic locality in the valley of Virginia. The first house he will probably reach is "Greenway Court," the residence, in the eighteenth century, of Lord Fairfax, who owned quite a third of the old State of Virginia, and at that day "Greenway Court" was habitually frequented by Washington, a mere youth, and surveyor of Lord Fairfax's countless acres.

The character of this old noble man was eccentric, and his life had been filled with romantic incidents. He was descended from an old Scotch-English Knight, Sir Thomas Fairfax, who lived at his estate called Denton, in Yorkshire. Failure of fortune, and bitter disappointment in a love affair, drove young Lord Fairfax into exile across the ocean. His early manhood had been brilliant; he had been educated at Oxford, was a member of the "Blues," and led the life of a fine London gentleman of the first water, in the midst of nobles, countesses and authors. The moment came when young Fairfax found himself entangled in one of those affairs which shape the destinies of men. He fell in love with a beauty of the court, paid his addresses to her, and was engaged to be married; every preparation was made, coaches, horses, jewels, costly presents of every description were ordered, and the blissful moment was near at hand—it was not fated to arrive. The young lady suddenly changed her mind, a ducal coronet was held up before her, by a rival, and she jilted young Fairfax. He became a bitter cynic and woman-hater thenceforth to the day of his death. His taste for cultivated and refined society forsook him, and bade adieu to England and crossed the ocean to his possessions in America, to the valley of the Shenandoah, then filled with deer and wolves, and buried himself in the vast wilderness where "Fans never flirted, nor Ribbons fluttered."

The lands mentioned were of princely extent and were inherited by him from his mother, a daughter of Lord Culpepper. They embraced the whole area lying between the Rappahannock and Potomac rivers in the colony of Virginia, from the shores of a certain Chesapeake Bay to the head-waters of the said rivers. Here was a new world of good fortune opened. Denton, Nun-Appleton, and all the English estates of the Fairfaxes, might have been hidden away in one corner of the Virginia principality, and lost from view. Rivers, bays, mountains, rich lowlands, breezy uplands, forests, mines, towns, and wild beasts in myriads to hunt, were the young lord's; and he was duke, prince, king almost, in the extent of his possessions. It is true that the country was comparatively unexplored; but settlers were thronging in; the ax of the pioneer was ringing in the great forests; fertile fields were coming steadily under cultivation; Fredericksburg, Winchester, Warrenton, and numerous other towns were springing up—of all which the bankrupt young earl found himself suzerain by letters patent from our lord the king, with rental of a shilling only, for each thousand or hundred thousand acres, payable each year at the "Feast of St. Michael the Archangel."

It was this handsome little property that young Thomas Lord Fairfax, Baron of Cameron, now came to visit. William Fairfax, a kinsman, had preceded him, and built a commodious residence for himself and family, called Belvoir, which was situated on the Potomac, below the present city of Washington. The family at Belvoir had many agreeable neighbors, among them the Washington family, living at Mount Vernon, near by. With the Washingtons, indeed, William Fairfax was connected by marriage; and, when Lord Fairfax came, he made their acquaintance, from which sprung his connection with young George Washington, then an unknown youth of fourteen or fifteen. It may be fairly said that the influence thus brought to bear upon the life of the boy had a paramount part in shaping his subsequent career. The youth was of an ardent and energetic temperament, and had longed to enter the royal navy, in which he had secured a midshipman's warrent—only to desist, however, from his intention in consequence of the tears and entreaties of his mother. He was now without occupation, and was idly spending his time in social visiting and in hunting. At this crisis Lord Fairfax appeared, made a favorite of the youth, conversed with him of England and the great world, and ended by engaging him to proceed across the Blue Ridge and survey the Fairfax lands toward the upper waters of the Potomac.

Young George Washington, then sixteen, ardently accepted this offer of the old nobleman, and in March, 1748, set out on horseback for the Blue Ridge, crossed at Ashby's Gap, and entered the valley—a stalwart, ruddy, manly young fellow, keen in quest of incident and adventure, and highly pleased, as he intimates, at the prospect of earning a "doubloon a day" as surveyor. The figure of the youth on his spirited horse, with chain and compasses and other instruments, rifle in hand, and a smile upon his lips, crossing the mountain, fording the Sheuandoah, and riding on beneath the great sycamores into the far-reaching prairies—this figure will forcibly arrest the attention of every student of the life of Washington. On the night of the day when he passed the Blue Ridge, he slept, he says in his journal, at "Lord Fairfax's;" and the spot thus designated was Greenway Court, to which Lord Fairfax soon afterward removed, making it thenceforward his chief place of residence.

The house of Greenway Court was situated near the present village of White Post, so called from a white post erected at the spot by Lord Fairfax, to point out the way to his dwelling.* It stood some miles from the Shenandoah, in the midst of a lovely country, which—beautiful to-day—is said to have been far more so in the last century. The English traveler Burnaby, journeying at that time from the east to the Blue Ridge, declares that, from the summit of the mountain, the exquisite landscape, brilliant with "camœ-daphnes in full bloom," burst on him like a fairy spectacle; and he exclaims that only to live here, poor and humble, were better than to be prince or king elsewhere! The valley of the Shenandoah is, indeed, a region of the rarest attractions. The beauties of the lowland and the mountains are blended in the landscape; on the left the Blue Ridge rolls away in azure billows; southward, the "Three Sisters" and the Massinutton rise like a battlemented wall, deep blue against the orange flush of sunset; to the west the great North Mountain stretches like a cloud along the horizon; and through fertile fields, or tall forests, the Shenandoah, limpid as crystal, steals, with a low murmur, by the base of the Ridge toward the Potomac. This country is noted now for its rich crops of cereals; in the last century it was equally famous for its luxuriant grass. Tradition declares that the valley at that time was one vast prairie, alternating with forest; that in summer the grass was so tall as to be tied together in front of a man on horseback; and that the prairie, extending as far as the eye could see, was dazzling, with its myriads of flowers in full bloom—an ocean of the richest colors, which every breeze broke into billows.

A more appropriate place of residence could not be imagined for the exiled nobleman aud disappointed lover, whom "man delighted not, nor woman either!"

*This post still stands, or rather a similar one, the authorities of the village carefully replacing it when it decays or is injured.

The old house of Greenway Court was torn down by the owner of the property many years since, but the writer saw it standing, and well remembers it. It was a long, low mansion, built of stone, with a veranda in front, overshadowed by ancient locust-trees, old dormer-windows lighting the attic, and two belfries on the roof, intended, it has been supposed, to contain bells for the purpose of calling together the settlers in case of an Indian attack. At the distance of fifty or a hundred yards, was a low stone-cabin, originally occupied by Lord Fairfax as an "office" for the issue of deeds to settlers, one of which, on yellow parchment, with the brief signature "Fairfax," is now before the writer. The larger "court" is said to have been designed by Lord Fairfax for his steward, which would seem to indicate an intention of erecting a more suitable mansion for himself. But he was an eccentric personage; seems to have preferred a small, wooden cabin near; and here, surrounded by his deer, hounds, English servants, rude retainers, and a few books, of which a list remains, he passed a number of the latter years of his life, a Virginia Nimrod, and king of the wilds.

Of the place and its figures, at that epoch, an idle fancy might draw an animated picture—a great fire blazing on the hearth of the small house, the autumn foliage brushing the roof, hounds sleeping on the floor or gamboling in the sunshine, and his lordship, the Earl of Fairfax, and Baron of Cameron, enthroned in hunting-garb in his great chair, surrounded by huntsmen, trappers, Indians—all the rude society, in a word, of the frontier. His passion was hunting, and it is said that he had also an eccentric fancy for hoarding English gold-coin—a considerable quantity was unearthed at the spot some years ago.

Tall, swarthy, reserved, and with no adjuncts of place or power, his lordship, nevertheless, preserved considerable state and dignity, it is said, as lieutenant of the county and chief magistrate, riding to court in his chariot drawn by four horses, and grandly presiding, wrapped in a rich red velvet cloak. Thus, in reading, hunting, dreaming, passed the long years of Fairfax's exile at Greenway Court; and the boy George Washington came and went, growing to manhood. At last the Revolution came, and the boy surveyor was appointed commander-in-chief of the American armies. What Lord Fairfax thought thereof is not known; but one last incident connects him with the ruddy boy. In 1781 the Earl was at Winchester, when a sudden commotion seized upon the people, and when he inquired its meaning, he was informed that Lord Cornwallis had surrendered his army at Yorktown to General George Washington. At this intelligence the aged Earl stood aghast. The boy to whom he had paid a doubloon a day for surveying, had annihilated the British dominion on the western continent. The old Earl uttered a a groan and exclaimed to his old body servant, "Take me to bed, Joe; it is time for me to die," and in a few months he did die.

A short distance from Greenway Court, and about two miles from the little villages of Millwood and White Post, each, is Saratoga, the ancient residence of the renowned victor of Tarleton, General Daniel Morgan. Saratoga is plain, massive, unpretending, embodying the character of its owner. Morgan, in his youth, fought the Indians about Winchester, defended Edward's Fort on Lost River against them, and in 1756 took part in Braddock's fatal expedition as a common soldier. In this campaign he received a bullet through the neck, and four hundred and ninety-nine lashes. Soon the Revolution came. He raised a company of the finest youths in Frederick, and a battalion in the Valley, and marched to join Washington at Boston. These were the first troops that marched from the South to the defense of the North. Morgan, on reaching Boston, drew up his Virginians, and at Washington's appearance, made the military salute and reported "from the right bank of the Potomac, General." The face of Washington flushed, his eyes filled, and dismounting, passed along the entire line, grasping every hand in turn.

Of Morgan's ability as a soldier there can be no doubt, and his merit in the campaigns of the Carolinas is fully recognized by General Greene and Colonel "Light-Horse Harry" Lee. At the Cowpens, the backwoodsman overthrew the brilliant Colonel Tarleton, trained in all the military science of the European school, and the result at the battle of Saratoga was claimed by his friends to have been largely due to his nerve and soldiership. He is said to have named his house "Saratoga," in grim, historic protest against the injustice of General Gates, who scarcely mentioned him in his bulletin of the battle. After the war, Morgan retired to the Valley, and erected this mansion—taking no part in public affairs thereafter, save once, as member of Congress from Frederick county.

Of this stalwart soldier—a tall, powerful, bony, and plain-spoken man—as of the building of his house, many traditions remain in the neighborhood. At Winchester, some miles distant, were stationed a large number of Hessians, taken prisoners at Saratoga, and as these men were, many of them, stone-masons by trade, Morgan employed them to build his house. The stone for the purpose, which is in large blocks, was quarried on the Opequon, and the Hessians are said to have borne it for miles on their shoulders, the General riding beside them, and spurring them on with the statement that, "If they did not work, the country could not afford to feed them!" Whether this be true or not, the General succeeded in constructing an excellent dwelling-house, and here were spent his calm, latter years. It is said that in process of time he became deeply pious, uniting himself to the Presbyterian Church; but, according to his own statement in his last days, he had always experienced strong religious impressions.

"People thought," he said on his death-bed, "that Daniel Morgan never prayed—they said that old Morgan never was afraid—they did not know. Old Morgan was often miserably afraid!"

"On the night," he declared, "of the storming of Quebec, in the deep darkness, he felt his heart sink, and, going aside, knelt down by one of the cannon, and prayed that the Lord God Almighty would be his shield and defence." In like manner, at the Cowpens, the sight of Tarleton's imposing forces in his front had filled him with dismay; whereupon he retired to the woods near at hand, and, kneeling in an old tree-top, prayed earnestly for himself, his men, and his country. This is assuredly the true spirit of the Christian warrior, shrinking, it may be, from death and judgment, but bravely doing his duty after prayer to God; and "Old Morgan" here presents a nobler spectacle than any whiskered "army-man" that ever swore to hide his trepidation.

That the zest of life, however, was powerful in this strong organization, there is every reason to believe. Physical health and strength made him enjoy life keenly, and relax his hold upon it with regret. A tradition remains, that on his death-bed, or in his latter days, he said to one of his friends:

"To be only twenty again, I would be willing to be stripped naked, and hunted through the Blue Ridge with wild dogs!"

Morgan died in Winchester in 1802, at the age of sixty-seven, but he lived until 1800 at the house of Saratoga. A visit to the place will repay the lover of historic localities. With its great dining-room, lofty mantel-pieces, decorated with bead-work and panneling, its elaborate wainscoting and ponderous walls resembling those of some feudal castle, the antique building carries you back to a period when all things seem to have been more solid, substantial, and enduring, than at present. You fancy that the house reflects the character of the person who erected it—a plain, unassuming man, making no professions, but genuine, strong, and to be relied upon. If the traveler who journeys hither be a lover also of the picturesque, his taste will not remain ungratified. Saratoga stands on a gentle knoll, half surrounded by an amphitheatre of wooded hills. In front, across the rolling Valley, rise the blue battlements of the Ridge; a hundred yards away bubble up the bright waters of the beautiful fountain; and the wide-spreading willows, drooping their tassels in the stream, sigh dreamily in unison with the reverie in which the visitor may indulge.

THE HOUSE OF GENERAL LEE.

Near the little hamlet of Leetown, and in the angle formed by the waters of the Opequon and the Potomac, stand the houses once occupied by two famous soldiers, exiles both, and embittered by disgrace—General Charles Lee and Horatio Gates.

Truth is stranger than fiction. The adage is trite, but pithy and true. It was surely a singular and striking coincidence that these two men should have come hither within a few miles of each other, to rust out lives once crammed with exciting incident, and crowned with honors. Both were Englishmen, and soldiers of fortune. Both had been major-generals in the American army. Both had fallen into disgrace, and been suspended from their rank. Both were, even during their lives, lost from public memory, and the very resting-places of their bodies are now forgotten. The writer has never passed the small house occupied by Lee, during nearly the last quarter of the last century, without strongly realizing the great contrast between a dwelling so humble, and the career of the human being who made this his home—if such a man could possess *a home*—for so many years. Lee's life would furnish material for an exciting romance; and the character of the man himself was as singular as any imagined by writers of fiction. He was by birth of the English gentry—the son of General John Lee, of the British army, who married a daughter of Sir Henry Bunbury, Bart. Entering the army a mere boy, he took part in the French war in America; was adopted as a chief by the Mohawk Indians, at twenty-four, under the name of "Boiling Water," which accurately describes his impetuous character; was shot through the body at the battle of Ticonderoga, while shouting "Stand by me, my brave grenadiers!" nearly lost his life at the siege of Fort Niagara; sailed across Lake Erie, and pierced the wilderness to Fort Du Quesne, going thence a journey of seven hundred miles to Crown Point; descended the St. Lawrence and witnessed the surrender of Montreal; and two years afterward was fording the Tagus, in Portugal, and making a night attack at the point of the bayonet on the Castle of Villa Velha. The King of Portugal made him aid-de-camp and major-general, but the war ended, and Lee came back to England. He could not rest. Scarce more than thirty, he opened a violent broadside, with his vigorous pen, on the party in power, which drove him from the army, and made him a wanderer and soldier of fortune. Thenceforth his life became more than ever a romance. He repaired to the court of Frederick the Great, and had long talks with that famous autocrat. His next step was to offer his sword to Stanislaus Augustus, King of Poland, who made him his aid-de-camp, and admitted him to his table and his intimacy. Eternal movement was, however, a necessity of this man's blood. He set off for Constantinople; nearly perished from cold and hunger in the mountains of Bulgaria; and, in Turkey, was wellnigh swallowed up by an earthquake. Thence he passed back like a meteor to England; solicited employment without success; wrote new and more bitter attacks than before upon the ministry; returned to Poland; was made major-general there, and joined the Russian allies, and fought the Turks at Chotzim, retreating with the Cossacks, who were terribly cut up by the Turkish cavalry. This terminated the military career of Lee in Europe. He left the Polish service, traveled restlessly, tormented by gout and rheumatism, in Italy, Sicily, Malta, and elsewhere—and these years were signalized by new assaults upon the English ministers, so bitter and brilliant as to have convinced many persons that Lee was the author of "Junius."

In 1773 the restless and disappointed soldier turned his eyes toward America—whose cause he had defended long and ably—and in the

4

same year we find him at Mount Vernon, consulting with Washington, who received him with the consideration due to his military ability, and his reputation as a soldier. With Mrs. Washington he appears to have been less of a favorite. He is said to have tramped, followed by his pack of dogs, through the fine drawing-rooms—had these canine pets to sit by him at table—and to have conducted himself in a manner not calculated to secure the good graces of a neat Virginia housewife. Lee's character and manners were probably a more serious obstacle to his popularity with ladies. He was bitter, cynical, sarcastic, and, it would seem, careless of his person—a thin, lanky, angular human being at the best, not such as delights the feminine eye. He appears to have indulged throughout life a habit of sneering at everything; and, when he left Mount Vernon, the lady of the house no doubt rejoiced at his departure.

The famous soldier was warmly welcomed by Congress, made major-general, and seems to have aspired to the chief command. It was wisely withheld, and fell to Washington—and the Monmouth business followed. Lee ordered, or was charged with ordering, his corps to fall back in the heat of action. Washington rode toward him, through the smoke, raging, with flaming eyes, uttering imprecations almost; and, after the battle, Major-General Lee was court-martialed, found guilty, and deprived of his rank in the army.

So ended, suddenly, all the brilliant dreams of the soldier of fortune, who had, no doubt, looked forward to becoming sooner or later, generalissimo of the American armies. The blow seems to have well nigh paralyzed him, for he never again made an effort to attain military position in America or elsewhere. His sentiment toward Washington became bitter beyond words; and he retired in wrath and disgust to the small stone-house in the Valley, near the Opequon, of which I have made mention in the commencement of this article.

Here, General Charles Lee lived the life of a cynic and full-blooded Diogenes. The interior of the house had no partitions, being divided, by imaginary lines merely, into chamber, sitting-room, kitchen, etc.; and in this cabin, surrounded by his dogs, with his saddle thrown down in one corner, Lee vegetated year after year. His only companion was an Italian body-servant, Minghini, and he rarely visited any one save General Gates, who lived some miles distant. His bitterness, cynicism, and blasphemous contempt for everything sacred, are clearly shown by well-established tradition. His hounds were named after the Holy Trinity and the Twelve Apostles, and he left directions in his will that his body should not be buried "in any church or church-yard, or within a mile of any Presbyterian or Anabaptist meeting-house; for, since he had resided in this country, he had kept so much bad company when living, that he did not choose to continue it when dead." When on a visit once to General Gates, a quarrel is said to have taken place between the latter and Mrs. Gates, who passionately demanded of General Lee his opinion on the merits of the controversy, and of herself.

This unlucky question gave Lee an opportunity to display all his Junius-like spleen. "Madam," he said, with mock ceremony and a bitter sneer, "my opinion of you is, that you are—*a tragedy in private life, and a farce to all the world!*"

With Washington, his relations remained embittered, and he wrote and published "Queries: Political and Military," in which he made a fierce attack on the great soldier. In after-years, it is said that Washington forgave or forgot these old enmities, and, when once in the Valley, sent word to General Lee that he would on a certain day come and dine with him. Lee's action was prompt. He mounted his horse and rode away. When Washington reached the house, he found tacked upon the front door a slip of paper containing the words, "*No meat cooked here to-day!*"

These incidents are given on the authority of neighborhood tradition. The general estimate of Lee is based, however, upon an old volume entitled "Memoirs of the Life of the late Charles Lee, Esq., Lieutenant-Colonel of the Forty-fourth Regiment, Colonel in the Portuguese Service, Major-General and Aid-de-Camp to the King of Portugal, and Second in Command in the Service of the United States of America, during the Revolution. London, 1792." It is possible that the writer was politically hostile to Lee, but there seems little reason to question the intense cynicism and bitterness of the soldier's character. After all, however, he was his own worst enemy. To his savage "Queries," Washington made no reply; and he sank into obscurity and utter neglect, which most of all must have galled his proud and arrogant nature. Nobody seemed to think him worth the trouble of notice. He growled in his solitary den in the wilderness, but his growls were unheeded. He would no doubt have died here, but on a visit to Philadelphia he was "seized with a shivering," and taking to his bed in an obscure inn called "The Conestoga Wagoner," never again rose. His last hours were passed in delirious mutterings, which indicated that his memory had returned to adventurous incidents of his career in Europe and America. The words uttered by him to his men at Ticonderoga were the last on his lips.

"Stand by me, my brave grenadiers!" he exclaimed. Soon after this fierce cry, the bitter exile expired.

He had been aid-de-camp and friend of kings, second in command in a republic, a writer so famous as to be thought the real Junius, and he died thus in the western wilds, lost from sight and memory. The traveler, passing the small, stone-house with its dilapidated enclosure, can scarcely realize that here dragged out the last years of a soldier and political writer once so famous.

"Traveler's Rest," the residence of General Gates, comes next into view. The singular coincidence in the lives of Lee and Gates was remarkable, and being neighbors intensifies one's interest in these two remarkable men. Gen. Gates was the son of Capt. Gates of the British Army, and Horace Walpole officiated as god-father at his christening. Entering the Royal American forces, he served in various quarters; gained credit in Martinque, was with Braddock on his expedition in 1756, to Fort Du Quesne, and returned to London. He again repaired to America in 1773. Went to Mount Vernon, where Lee then was, and duly received a high commission in the army. At the battle of Camden his ambition was overthrown, Congress deprived him of his command, and he retired to the Valley of Virginia, purchased a house, and then spent his days in retirement and obscurity. The house occupied by him bears the name of "Traveler's Rest." Here General Horatio Gates had once glittered in the zenith of fame, here he dragged out his latter days in obscurity.

GUIDE

TO

BALTIMORE & OHIO RAILROAD.

1,000 Copies of the "Guide" will be issued Monthly, and be for sale upon every Passenger Train of the Baltimore & Ohio Railroad, Baltimore & Washington Railroad, and Valley Railroad, and in all the Hotels, News Depots and Railway Stations in all the towns and cities from Boston to Cincinnati, St. Louis, Pittsburgh, Richmond and Chicago.

RATES OF ADVERTISING.

No Advertisement Inserted for Less than $5.

¼ Page, per Month.....	$ 5	00
½ do. do.	10	00
¾ do. do.	15	00
1 do. do.	20	00

Advertisements inserted on either the Top, Bottom or Outer Margin of every Page running through entire Guide, according to special contract.

Advertisements inserted for six months, an abatement of 10 per cent. on the above charges.

Advertisements inserted for twelve months, an abatement of 15 per cent. will be made on the above charges.

☞ All advertising charges will be cash on appearance of the advertisement in the "Guide."